INVERSE AND ILL POSED PROBLEMS SERIES

Nonclassical and Inverse Problems for Pseudoparabolic Equations

Other titles in the
Inverse and Ill-Posed Problems Series:

Methods for Solving Operator Equations
V.P. Tanana

Formulas in Inverse and Ill-Posed Problems
Yu.E. Anikonov

Inverse Logarithmic Potential Problem
V.G. Cherednichenko

Multidimensional Inverse and Ill-Posed Problems for Differential Equations
Yu.E. Anikonov

Ill-Posed Problems with A Priori Information
V.V. Vasin and A.L. Ageev

Integral Geometry of Tensor Fields
V.A. Sharafutdinov

Inverse Problems for Maxwell's Equations
V.G. Romanov and S.I. Kabanikhin

Related Journal:

Journal of Inverse and Ill-Posed Problems
Editor-in-Chief: M.M. Lavrent'ev

INVERSE AND ILL-POSED PROBLEMS SERIES

Nonclassical and Inverse Problems for Pseudoparabolic Equations

A. Asanov and E.R. Atamanov

///VSP///

Utrecht, The Netherlands, 1997

VSP BV
P.O. Box 346
3700 AH Zeist
The Netherlands

First published in 1997

ISBN 90-6764-235-5

Printed in The Netherlands by Ridderprint bv, Ridderkerk.

Contents

Introduction

In the monograph we have studied some nonclassical and inverse problems for parabolic equations describing the processes of transfer in composite media. We have considered the interior problems, i.e. the problems of the re-establishment of a solution by the given data in the domain of regularity (the probe problems). We have investigated also inverse problems for operator differential and integro-differential equations of the pseudoparabolic type. The considered problems, as a rule, are ill-posed in the Hadamard sense. We point out the classes of uniqueness for the considered problems. Furthermore, we obtain stability estimates and existence theorems for the solutions of the above-mentioned equations.

Different parts of the ill-posed problem theory may be referred to the traditional sections of mathematics. The problems considered in this monograph may be related to the theory of differential equations with partial derivatives from the ill-posed problem theory (see Lavrent'ev, 1962, 1977; Romanov, 1973 a,b, 1984; Tikhonov and Arsenin, 1974; Anikonov, 1978; Ivanov *et al.*,1978; Lavrent'ev *et al.*,1980; Tanana, 1981; Fedotov, 1982; Bukhgeim, 1983; Romanov *et al.*,1984).

Pseudoparabolic equations were probably first defined in describing wave propagation in heterogeneous media (Rubinstein, 1948). Then there were the papers of Barenblatt *et al.*, (1960), Barenblatt (1963), and Dzertser (1972), where a pseudoparabolic equation arises when liquid or gas filtration is simulated in porous media. The importance of investigating the problem of pseudoparabolic equations was mentioned by Lions (1972). One of the foremost publications - a classic work on pseudoparabolic equations - is that of Hal'per (1960). Henceforth, interest in these equations grew because of their numerous applications (Colton, 1972 a, b; Rao and Ting, 1972, 1973; Colton and Wimp, 1979; Rundel, 1979, 1980; Suvejka, 1980, 1983; Shkhanyukov, 1982; Vodakhova, 1984; Matveev and Sviridjuk, 1985; Sviridjuk, 1985).

The nonclassical problems for pseudoparabolic equations have been much less intensively investigated (Atamanov, 1977, 1979, 1983; Shishatskii and Atamanov, 1978; Shishatskii, 1980; Atamanov and Mamayusupov, 1990). Nevertheless, it is important to study these problems because of their practical utility and because some important ill-posed problems are considered briefly in well-known monographs.

The problem of continuation from the time-like half-line arises when it is required, from only indirect measurements, to find a physical field in all the domain. This problem is considered in the first chapter. The interior problems investigated in the second chapter may be obtained when stadying probe questions. The third chapter deals with problems of determining the coefficients and kernels of the operator integro-differential pseudoparabolic equations. Such equations describe the interior properties of a material, where physical and chemical parameters may change rapidly. Information about the solution of these equations is derived from measurements of such processes or parameters in the accessible part of the domain.

The problems considered in this monograph are generally ill-posed in the sense of Hadamard. The basic topic that is discussed is the conditional well-posedness of such problems. This represents the proof of uniqueness and stability theorems in the appropriate classes of functions. In some cases, existence theorems are also considered.

The results of Sections 1.1 and 1.2 were obtained by Shishatskii and Atamanov (1978); very similar to these were the results of Lavrent'ev *et al.*(1980). The results of Section 1.3 were obtained by S.P. Shishatskii. The results of Section 1.6 and 2.5 were not published by Shishatskii and are included in the monograph as a tribute to him.

The remaining material of Chapter 1 and 2 has been written by E.R. Atamanov. The results of Chapter 3 were obtained by A. Asanov and E.R. Atamanov.

The authors express their sincere gratitude to the outstanding mathematician Sergei Petrovich Shishatskii. It was his influence that stimulated our interest in nonclassical inverse problems and, as a result of which, the main results of the book were obtained.

Chapter 1

The Cauchy problem with data in a time-like manifold for the pseudoparabolic equation.

Here we shall consider the questions of uniqueness, stability and existence of the solution to the ill-posed by Hadamard problem - continuation of a solution to the pseudoparabolic equation from the time-like half-line. Uniqueness and stability for the Cauchy problem with data in a time-like manifold for different equations have been considered by many authors. For parabolic equations, the problem with data in a time-like manifold was first considered by Holmgren (1905). Further, uniqueness for the more general one-dimensional parabolic equation has been considered by Landis (1959); he also obtained the stability estimate for the solution. The uniqueness theorem for a solution to the Cauchy problem with data in a time-like surface for second-order parabolic equation in more general cases was almost simultaneously obtained by Shirota (1957), Mizohata (1959), and Protter (1961). The estimate for solutions to these problems was derived by Amonov and Shishatskii (1972).

Uniqueness of solution to the Cauchy problem with data in a time-like manifold for hyperbolic equations was investigated by Holmgren (1905) and Nirenberg (1957). The general uniqueness theorems were proved by Calderon (1958), Hörmander (1965), and Shishatskii (1973). Stability estimates for the corresponding problems were obtained by John (1960), Garipov (1974), Garipov and Kardakov (1973), and Kaistrenko (1975).

In the case of analytic coefficients, the uniqueness theorem for the Cauchy problem solution to elliptic equations is the corollary of the general Holmgren theorem (see Holmgren, 1905). For the case of two independent variables, the Cauchy problem uniqueness was proved by Carleman (1939). The uniqueness theorem for the solution to the Cauchy problem for general elliptic equations was considered by Douglis (1953), Hartman and Winther (1953), Cordes (1960), Lavrent'ev (1956), and Landis (1963). The stability of the Cauchy problem solution to the second-order elliptic equation considered in the plane for the class of bounded solutions was proved by Carleman (1933). The estimates of the solution's stability for the Cauchy problem in the considered domain were obtained by Lavrent'ev (1962), Landis (1963), and others in the case of arbitrary linear elliptic equations with sufficiently smooth coefficients.

Here we consider the problem with data in the characteristic. When constructing the uniqueness classes for those problems, one usually assumes that the solution is determined in the half-space and satisfies certain additional conditions. Note that in our case, we may obtain the uniqueness classes for solutions determined only in the quarter-space.

So, consider the Cauchy problem

$$u(0,t) = f_1(t), u_x(0,t) = f_2(t) \tag{1.0.1}$$

for the parabolic equation

$$u_t(x,t) = u_{xxt}(x,t) + u_{xx}(x,t). \tag{1.0.2}$$

Since the line $x = 0$ is a characteristic, then in the case when $f_1(t), f_2(t)$ are given locally (i.e., in the compact set) the solution to this problem is not unique (see Rundell and Stecher, 1977).

§1.1 Uniqueness of the solution in the class $U(a,b)$

We assume that the solution to problem (1.0.1), (1.0.2) is determined in the quadrant

$$P = \{0 \le x < \infty, 0 \le t < \infty\},$$

and in a certain class of solutions we shall prove the uniqueness theorem.

Definition. We say that the solution $u(x,t)$ of equation (1.0.2) belongs to the class $U(a,b)$ if the functions u, u_γ are continuous in P and

$$|u|, |u_\gamma| \le C_u e^{ax+bt},$$

where $\gamma = x, t, xx, xt, xxt$. The numbers a, b satisfy the following conditions

1) if $a < 1$ then b is real,

2) if $a \ge 1$ then $b < -1 + \left[4a(a+\sqrt{a^2+1})\right]^{-1}$.

C_u is a constant depending on u.

The class $U(a,b)$ is not empty. Really, let, for example, $a > 1$. It is easy to show that for a real α the function

$$u(x,t) = \operatorname{Re}\exp\left\{(a+i\alpha)x + \left[-1 + \frac{1-a^2+\alpha^2}{1-a^2+\alpha^2+4a^2\alpha^2} + \frac{2a\alpha}{(1-a^2+\alpha^2)^2+4a^2\alpha^2}i\right]t\right\}$$

solves the equation (1.0.2). If

$$\alpha \ne \pm\sqrt{a^2-1+2a\sqrt{a^2-1}},$$

then the inequality

$$-1+\frac{1-a^2+\alpha^2}{(1-a^2+\alpha^2)^2+4a^2\alpha^2} < -1+[4a(a+\sqrt{a^2-1})]^{-1}$$

holds, i.e.

$$u(x,t) \in u\left[a, -1+\frac{1-a^2+\alpha^2}{(1-a^2+\alpha^2)^2+4a^2\alpha^2}\right].$$

Note, that for $\alpha \to \pm\sqrt{a^2-1-2a\sqrt{a^2-1}}$ we have

$$-1+\frac{1-a^2+\alpha^2}{(1-a^2+\alpha^2)^2+4a^2\alpha^2} \to -1+[4a(a+\sqrt{a^2-1})]^{-1}.$$

Lemma. Let

$$f(P,S)=\frac{1-P^2+S^2}{(1-P^2+S^2)^2+4P^2S^2}.$$

Then

$$\sup_{p\geq a, -\infty<S<\infty} = \begin{cases} \left[4a(a+\sqrt{a^2-1})\right] & \text{for} \quad a\geq 1, \\ \infty & \text{for} \quad a\geq 1, \end{cases}$$

Proof. The proof implies from the maximum principle in the harmonic half-plane Re$z > a$ for the function

$$f(P,S)=\text{Re}(1-z^2)^{-1}, z=P+iS.$$

If $a\geq 1$, then $f(P,S)$ achieves the maximum in the line Re$z=a$; if $a<1$, then $f(P,S)$ achieves the maximum in the point $z=1$. In this case, $\sup f(P,S)=\pm\infty$.

Not too complicated calculations show that for $a\geq 1$

$$\max f(a,S)=f(a,\pm\sqrt{a^2-1+2a\sqrt{a^2-1}})=[4a(a+\sqrt{a^2-1})]^{-1}.$$

Theorem 1.1. Problem (1.0.1), (1.0.2) has no more than one solution from the class $U(a,b)$.

Proof. We show that from the conditions

$$f_1(t) = f_2(t) = 0 \quad (t \geq 0), \quad u \in U(a,b) \tag{1.1.1}$$

it follows that $u(x,t) \equiv 0$. Let

$$z = P + iS, \quad \hat{u}(z,t) = \int_0^\infty e^{-zx} u(x,t)dx.$$

As $u(x,t) \in U(a,b)$, the function $\hat{u}(x,t)$ is analytic for $P > a$. Applying the Laplace transformation to equation (1.0.2) we obtain

$$\hat{u}_t(z,t) = -u_{tx}(0,t) - u_x(0,t) - zu_t(0,t)$$

$$- zu(0,t) + z^2 \left[\hat{u}_t(z,t) + \hat{u}(z,t)\right]. \tag{1.1.2}$$

This and (1.1.1) yield

$$(1 - z^2)\hat{u}_t(z,t) = z^2 \hat{u}(z,t),$$

whence, for $z \neq 1$, it follows that

$$\hat{u}(z,t) = C(z) \exp[z^2(1 - z^2)^{-1}t].$$

We now show that $C(z) \equiv 0$. The latter equality for $z \neq 0$ yields

$$C(z) = \exp[-z^2(1 - z^2)^{-1}t]\hat{u}(z,t) = \exp\{[1 - f(P,S)$$

$$- \frac{2PS}{(1 - P^2 + S^2)^2 + 4P^2S^2} i]t\} \int_0^\infty e^{-zx} u dx. \tag{1.1.3}$$

Whence it follows that

$$|C(z)| \le \exp\{[1 - f(P,S)]t\} \int_0^\infty e^{-Px} C_u e^{ax+bt} dx$$

$$= C_u (P-a)^{-1} \exp\{[1 - f(P,S) + b]t\}. \qquad (1.1.4)$$

The definition of the class $U(a,b)$ and the lemma yield

$$\inf_{P \ge a} [1 - f(P,S) + b] = 1 - \sup_{P \ge a} f(P,S) + b < 0.$$

It is easy to see that for each a, b from the half-plane $P > a$ there exists a continuous line $l, (1,0) \in l$, where

$$1 - f(P,S) + b < 0. \qquad (1.1.5)$$

In practice, for $a > 1$, l may be taken as any of the curves

$$S = \pm\sqrt{P^2 - 1 + 2P\sqrt{P^2 - 1}}, a + \delta \le P \le a + 2\delta.$$

If $a < 1$, l may be taken as the interval

$$S = 0, \quad 1 - 2\delta \le P \le 1 - \delta$$

(in both cases $\delta > 0$ is sufficiently small). Assuming in (1.1.2), $(P,S) \in l$ and $t \to \infty$ we obtain $C(z) = 0$ in l. Note, that for $u \in U(a,b)$ and $\text{Re}z > a$ we have $\hat{u}(z,t) \to \hat{u}(z,0)$ for $t \to 0$ and going to the limit for $t \to 0$ in (1.1.1) we obtain $C(z) = \hat{u}(z,0)$ for $\text{Re}z > a$, $z \ne 0$. Defining $C(z)$ in the point $z = 1$ (if $a < 1$) by the continuity, we obtain the analytic for $\text{Re}z > 0$ function, which vanishes in the curve l (the curve l is lying in this half-plane). Consequently, $C(z) \equiv 0$, whence $\hat{u}(z,t) \equiv 0$. As for the functions from $U(a,b)$ the Laplace transformation by x is uniquely reversible, then $u(x,t) \equiv 0$. The theorem is proved.

Let us show now that the uniqueness class $U(a,b)$ may be not extended essentially. Let for $a > 1$

$$u(x,t) = \frac{1}{2\pi i}\int\limits_{a-i\infty}^{a+i\infty} e^{zx}(z-1)^{-4}\exp[z^2(1-z^2)^{-1}t]dz$$

$$= \frac{1}{2\pi}e^{ax-1}\int\limits_{-\infty}^{\infty} e^{isx}(a+is-1)^{-4}\exp[f(a,s)$$

$$+\frac{2as}{(1-a^2+s^2)^2+4a^2s^2}i]tds.$$

Integrals obtained by differentiating twice by x and once by t the integrand function uniformly converge. Consequently, the derivatives of the function $u(x,t)$ from the definition of the class $U(a,b)$ exist and are continuous. The modulus of these derivatives may be estimated from above by the expression

$$C\exp[ax + (-1 + \max_s f(a,s))t]$$

$$= C\exp[ax + (-1 + \frac{1}{4a(a+\sqrt{a^2-1})})t].$$

It is easy to show that $u(x,t)$ solves equation (1.0.2). Finally, from the equalities

$$\lim_{z\to\infty} z^2\hat{u}(z,t) = \lim_{z\to\infty} z^2(z-1)^{-4}\exp[z^2(1-z^2)^{-1}t] = 0,$$

and from the well-known property of the Laplace transformation it follows that

$$\lim_{x\to 0} u(x,t) = \lim_{x\to 0} u_x(x,t) = 0.$$

Thus, $u(x,t)$ is nontrivial solution to the homogeneous Cauchy problem (1.0.1), (1.0.2).

Hence it follows that the inequality sign in the definition of parameter b (see definition of the class $U(a,b)$) could not be replaced by the equality sign with preservation of uniqueness.

§1.2 Stability of solutions in the class $U(a,b)$.

Solutions to the problem (1.0.1), (1.0.2) which belong to the classes $U(a,b)$ are unstable relative to the variations of the Cauchy data $f_1(t)$, $f_2(t)$. Let, for example, for $a > 1$

$$u^{(a)}(x,t) = a^{-2}\exp[ax - a^2(a^2-1)^{-1}t].$$

It is easy to verify that each function $u^{(\alpha)}(x,t)$ solves equation (1.0.2). Since for $a > 1$

$$-a^2(a^2-1)^{-1} = -1 - (a^2-1)^{-1} < -1 + [4a(a+\sqrt{a^2-1})]^{-1},$$

then

$$u^{(a)}(x,t) \in U(a, -a^2(a^2-1)^{-1}).$$

The Cauchy data for the solution $u^{(a)}(x,t)$ satisfy the estimate

$$|f_1(t)|, |f_1'(t)|, |f_2(t)|, |f_2'(t)| \leq \frac{a}{a^2-1}\exp(-\frac{a^2}{a^2-1}t) \to 0.$$

On the other hand, the solution $u^a(x,t)$ in each fixed point with $x > 0$ grows exponentially with a. This instability occures and for solutions which belong to the class $U(a,b)$. If, for example, for a natural k

$$u^{(k)}(x,t) = \exp\left[(-1+\frac{1}{k^2+1})t\right]\sin kx$$
$$-\frac{1}{2}\exp\left[(-1+\frac{1}{4k^2+1})t\right]\sin 2kx,$$

then $u^{(k)}(x,t)$ solves (1.0.2) and belongs to the class $u(0, -1+(k^2+1)^{-1})$. It is easy to show that for the functions

$$f_1(t) = u(0,t) = 0, f_2(t) = u_x(0,t)$$

$$= k\left\{\exp\left[(-1+\frac{1}{k^2+1})t\right] - \exp\left[(-1+\frac{1}{4k^2+1})t\right]\right\}$$

the following estimates (C is an absolute constant)

$$||f_1||_{C^n[0,\infty)}, ||f_2||_{C^n[0,\infty)} \le C^n k^{-1} \to 0, k \to \infty$$

hold. At the same time

$$u^{(k)}(\frac{\pi}{2k}, t) = \exp\left[(-1 + \frac{1}{4k^2+1})t\right] \sin\frac{\pi}{2}$$

$$-\frac{1}{2}\exp\left[(-1 + \frac{1}{4k^2+1})t\right] \sin\pi = \exp\left[(-1 + \frac{1}{k^2+1})t\right]$$

is not small.

We now show that for the solutions from the fixed class $U(a,b)$ and such that $C_u \le C$, where C is an absolute constant, the continuous dependence of the solution from the Cauchy data (1.0.1) variations holds true.

Theorem 2.2. Let for $u(x,t) \in U(a,b)$ which solves (1.0.1), (1.0.2) let the following conditions hold:

$$C_u = 1; |f_1(t)|, |f_1'(t)|, |f_2(t)|, |f_2'(t)| \le \varepsilon e^{bt}, 0 \le t < \infty, \qquad (1.2.1)$$

where ε is sufficiently small. Then, there exist $p_1 > a$, $C = C(a,b)$, $\alpha = \alpha(a,b,p_1) > 0$, such that

$$|u(x,t)| \le C\left(\frac{\ln\ln\varepsilon^{-\alpha}}{\ln\varepsilon^{-\alpha}}\right)^{1/2} e^{p_1 x + bt}.$$

Proof. Due to the linearity of the problem, the condition $C_u = 1$ is not essential.

Conditions (1.1.2), (1.0.1) yield

$$(1-z^2)\hat{u}_t(z,t) - z^2\hat{u}(z,t) + z\left[f_1(t) + f_1'(t)\right]$$

$$+f_2(t) + f_2'(t) = 0.$$

Integrating this equation and using for the arbitrary constant the notation $C(z)$ for $\mathrm{Re}z > a$, $z \neq 1$, we obtain

$$\hat{u}(z,t) = \exp\left(\frac{z^2}{1-z^2}t\right)\left\{C(z) - \int_0^t \left[\frac{f_1(\tau) - f_1'(\tau)}{1-z^2}z \right.\right.$$
$$\left.\left. + \frac{f_2(\tau) + f_2'(\tau)}{1-z^2}\right] \exp\left(-\frac{z^2}{1-z^2}\tau\right) d\tau\right\}. \quad (1.2.2)$$

It turns out that, under our conditions, the function $C(z)$ may be expressed by means of the given functions $f_1(t)$ and $f_2(t)$. For $z \in l$ (see Theorem 1.1, proof) for $t \to \infty$ and using (1.2.2) we obtain.

$$\left|c(z) - \int_0^1 \left[\frac{f_1(\tau) + f_1'(\tau)}{1-z^2}z + \frac{f_2(\tau) + f_2'(\tau)}{1-z^2}\right] \exp\left(-\frac{z^2\tau}{1-z^2}\right) d\tau\right|$$
$$= \left|\exp\left(-\frac{z^2}{1-z^2}t\right)\hat{u}(z,t)\right| = \left|\exp\left(-\frac{z^2}{1-z^2}t\right)\int_0^\infty e^{-zx}u(x,t)dx\right|$$
$$\leq \exp(-\mathrm{Re}\frac{z^2}{1-z^2}t)\int_0^\infty e^{-px}e^{az+bt}dx$$
$$= \frac{1}{p-a}\exp\{[1 - f(p,s) + b]t\} \to 0.$$

Consequently, for $z \in l$ we have

$$C(z) = \int_0^\infty \left[\frac{f_1(\tau) - f_1'(\tau)}{1-z^2}z + \frac{f_2(\tau) + f_2'(\tau)}{1-z^2}\right] \exp\left(-\frac{z^2x}{1-z^2}\right) d\tau,$$

and thus, using (1.2.2), for $z \in l$ we obtain

$$\hat{u}(z,t) = \exp\left(\frac{z^2t}{1-z^2}\right)\int_t^\infty \left[\frac{f_1(\tau) + f_1'(\tau)}{1-z^2}z \right.$$
$$\left. + \frac{f_2(\tau) + f_2'(\tau)}{1-z^2}\right] \exp\left(-\frac{z^2}{1-z^2}\tau\right) d\tau. \quad (1.2.3)$$

The function $f(P,S)$ is continuous in a certain neighborhood of the curve l and satisfies in l inequality (1.1.1). It is easy to show that there exist $\theta > 0$ and the interval L which lies in the domain $\mathrm{Re}z > a$ and is parallel to the line $\mathrm{Re}z = a$, such that for $r \in L$ the inequality

$$\mathrm{Re}[-z^2(1-z^2)^{-1}] + b = 1 - f(P,S) + b < -\theta$$

holds. Note, that $\theta = \theta(a,b)$, $L = L(a,b,\theta)$. If $a > 1$, then the interval L may be chosen arbitrarily close to the line $\mathrm{Re}z = 0$; the length of this interval depends on a, b and θ. If $a < 1$, then, generally speaking, it is necessary to choose the interval L close to the point $z = 1$, and its length in this case depends on b and θ.

For $z \in L$, using (1.2.1) and (1.2.3), we obtain

$$|\hat{u}(z,t)| \leq 2\varepsilon \exp(\mathrm{Re}\frac{z^2}{1-z^2}t) \int_t^\infty \left(\left|\frac{z^2}{1-z^2}\right| + \left|\frac{1}{1-z^2}\right| \right)$$

$$\times \exp\left[(\mathrm{Re}\frac{-z^2}{1-z^2} + b)\tau\right] d\tau \leq C(a,b)\varepsilon \left(\mathrm{Re}\frac{z^2}{1-z^2} - b\right)^{-1}$$

$$\times \exp\left[(\mathrm{Re}\frac{z^2}{1-z^2} + b)t\right] \leq C(a,b)\theta^{-1}\varepsilon e^{bt}.$$

So, for $z \in L \subset \{z : \mathrm{Re}z > a\}$, the inequality

$$|\hat{u}(z,t)| \leq C(a,b)\varepsilon e^{bt} \tag{1.2.4}$$

holds. (Here and later we shall use the notation C to denote any positive constant).

Note, that in the half-plane $\mathrm{Re}z > a$ for the function $\hat{u}(z,t)$, the estimate

$$|\hat{u}(z,t)| = \left|\int_0^\infty e^{-zx} u(x,t)dx\right| \leq \int_0^\infty e^{-px+ax+bt}dx = \frac{1}{p-a}e^{bt} \tag{1.2.5}$$

holds. Let $w(z,L)$ be the harmonic measure of the interval L relative to the half-plane z. Further, we shall need the estimate from above of the value $\min w(z,L)$ in the interval $\mathrm{Re}z = p_1$, $|Jmz| \leq \lambda$ for the big λ. Remember that $w(z,L)$ is the angle divided by π under which the interval L is seen from the point z. If $L = \{z : \mathrm{Re}z = p_0, |Jmz - z_0| \leq \Delta\}$, then

$$w(z,L) = \frac{1}{\pi}\tan^{-1}\left[\frac{2\Delta(\mathrm{Re}z - p_0)}{(\mathrm{Re}z - p_0)^2 + |(Jmz - s_0)^2 - \Delta^2|}\right].$$

Hence for $p_1 > p_0$ and λ sufficiently large

$$\min w(z,L) = \frac{1}{\pi}\tan^{-1}\left[\frac{2\Delta(p_1 - p_0)}{(p_1 - p_0)^2 + \lambda^2 - \Delta^2}\right]$$

$$> \frac{1}{\pi}\tan^{-1}\left[\frac{2\Delta(p_1 - p_0)}{(p_1 - p_0)^2 + \lambda^2}\right] > \frac{\Delta(p_1 - p_0)}{\pi\lambda^2}. \tag{1.2.6}$$

holds true. Applying to the half-space $\mathrm{Re}z \geq p_0$ the Nevanlinna theorem (see Evgrafov, 1968) about two constants and using (1.2.4)-(1.2.6) in the interval $\mathrm{Re}z = p_1$, $|\mathrm{Im}z| \leq \lambda$, we obtain

$$|\hat{u}(z,t)| \leq \left[C(a,b)\varepsilon e^{bt}\right]^{w(z,L)}\left(\frac{e^{bt}}{p_0 - a}\right)^{1-w(z,L)}$$

$$= C^w(a,b)(p_0 - a)^{-1+w}\varepsilon^w e^{bt} \leq \frac{C(a,b)}{p_0 - a}\exp\left[\frac{\Delta(p_1 - p_0)}{\pi\Lambda^2}\ln\varepsilon + bt\right].$$

We may assume that b in inequalities (1.2.4), (1.2.5) is such that $C(a,b) > 1$, $(p_0 - a)^{-1} > 1$. Hence

$$\left|\frac{1}{2\pi i}\int\limits_{p_1 - i\lambda}^{p_1 + i\lambda} e^{zx}\hat{u}(z,t)dt\right| \leq \frac{\lambda C(a,b)}{p_0 - a}\exp\left[\frac{\Delta(p_1 - p_0)}{\pi\lambda^2} \in \varepsilon + p_1 z + bt\right] \tag{1.2.7}$$

holds.

Integrating the inequality

$$\hat{u}(z,t) = \int_0^\infty e^{-zx} u(x,t)dx$$

by parts and using (1.0.1) we have

$$\hat{u}(z,t) = \frac{1}{z} f_1(t) + \frac{1}{z^2} f_2(t) + \frac{1}{z^2} \int_0^\infty e^{-zx} u_{xx}(x,t)dx. \tag{1.2.8}$$

Taking into account (1.2.1) we obtain

$$\left| \frac{f_1(t)}{2\pi i} \left(\int_{P_1 - i\infty}^{P_1 + i\infty} + \int_{P_1 + i\lambda}^{P_1 + i\infty} \right) e^{zx} \frac{dz}{z} \right|$$

$$= \frac{|f_1(t)|}{2\pi} e^{p_1 x} \left| \left(\int_{-\infty}^{-\lambda} + \int_{\lambda}^{\infty} \right) \frac{\cos sx + i \sin sx}{p_1 + is} ds \right|$$

$$= \frac{|f_1(t)|}{2\pi i} e^{P_1 x} \left| \left(\int_{-\infty}^{-\lambda} + \int_{\lambda}^{\infty} \right) \right.$$

$$\left. \times \frac{p_1 \cos sx + s \sin sx + i(-s \cos sx + p_1 \sin sx)}{p_1^2 + s^2} ds \right|$$

$$= \frac{|f_1(t)|}{2\pi} e^{p_1 x} \left| \left(\int_{-\infty}^{-\lambda} + \int_{\lambda}^{\infty} \right) \frac{p_1 \cos sx + s \sin sx}{p_1^2 + s^2} ds \right| \le C \varepsilon e^{p_1 x + bt} \tag{1.2.9}$$

with a certain absolute constant C (recall that we consider λ sufficiently large). Analogously,

$$\left| \frac{f_2(t)}{2\pi i} \left(\int_{p_1 - i\infty}^{p_1 - i\lambda} + \int_{p_1 - i\lambda}^{p_1 + i\infty} \right) e^{zx} \frac{dz}{z^2} \right| \le C \varepsilon e^{p_1 x + bt}. \tag{1.2.10}$$

Further,

$$\left|\frac{1}{2\pi i}\left(\int\limits_{p_1-i\infty}^{p_1-i\lambda}+\int\limits_{p_1-i\lambda}^{p_1+i\infty}\right)\frac{e^{zx}}{z^2}\int\limits_0^\infty e^{-zx}u_{\xi\xi}(\xi,t)d\xi dz\right|$$

$$\leq\frac{1}{\pi}\int\limits_\lambda^\infty\frac{e^{p_1x}}{p_1^2+s^2}\int\limits_0^\infty e^{-4\xi+a\xi+bt}d\xi ds=\frac{1}{\pi(p_1-a)}e^{p_1x+bt}\int\limits_\lambda^\infty\frac{ds}{p_1^2+s^2}$$

$$<\frac{1}{\pi(p_1-a)}e^{p_1x+bt}\int\limits_\lambda^\infty\frac{ds}{s^2}=\frac{1}{\pi(p_1-a)}\frac{1}{\lambda}e^{p_1x+bt}. \tag{1.2.11}$$

Formulae (1.2.7)-(1.2.11) yield

$$|u(z,t)|=\left|\frac{1}{2\pi i}\int\limits_{p_1-i\infty}^{p_1+i\infty}e^{zx}\hat{u}(z,t)dz\right|\leq\left|\frac{1}{2\pi i}\int\limits_{p_1-i\lambda}^{p_1+i\lambda}e^{zx}\hat{u}(z,t)dz\right|$$

$$+\left|\frac{1}{2\pi i}\left(\int\limits_{p_1-i\infty}^{p_1-i\lambda}+\int\limits_{p_0+i\lambda}^{p_0+i\infty}\right)e^{zx}\hat{u}(z,t)dz\right|$$

$$<\left\{\frac{C(a,b)\lambda}{p_0-a}\exp\left[\frac{\Delta(p_1-p_0)}{\pi\lambda^2}\ln\varepsilon\right]+2C\varepsilon+\frac{1}{\pi(p_1-a)\lambda}\right\}e^{p_1x+bt}.$$

Increasing if it is necessary $C(a,b)$ and denoting for brevity $\Delta(p_1-p_0)\pi^{-1}=\alpha$, from the latter inequality we obtain

$$|u(x,t)|<\frac{C(a,b)}{p_0-a}\left[\exp\left(\frac{\alpha}{\lambda^2}\ln\varepsilon\right)\lambda+\frac{1}{\lambda}\right]e^{p_1x+bt}. \tag{1.2.12}$$

Set here

$$\lambda=\left(\frac{\ln\varepsilon^{-\alpha}}{\ln\ln\varepsilon^{-\alpha}}\right)^{1/2}. \tag{1.2.13}$$

The previous estimates were obtained only for sufficiently large λ; therefore, in order that (1.2.13) holds, ε has to be sufficiently small. Substituting (1.2.13) into (1.2.12) we obtain

$$|u(x,t)| < \frac{C(a,b)}{p_0 - a}\left[\exp\left(-\ln\ln\varepsilon^{-\alpha}\right)\left(\frac{\ln\varepsilon^{-\alpha}}{\ln\ln\varepsilon^{-\alpha}}\right)^{\frac{1}{2}}\right.$$

$$\left.+\left(\frac{\ln\ln\varepsilon^{-\alpha}}{\ln\varepsilon^{-\alpha}}\right)^{\frac{1}{2}}\right] e^{p_1x+bt} = \frac{C(a,b)}{p_0-a}\left[(\ln\varepsilon^{-\alpha}\ln\ln\varepsilon^{-\alpha})^{-\frac{1}{2}}\right.$$

$$\left.+\left(\frac{\ln\ln\varepsilon^{-\alpha}}{\ln\varepsilon^{-\alpha}}\right)^{\frac{1}{2}}\right] e^{p_1x+bt} < C\left(\frac{\ln\ln\varepsilon^{-\alpha}}{\ln\varepsilon^{-\alpha}}\right)^{\frac{1}{2}} e^{p_1x+bt} \qquad (1.2.14)$$

with a certain new constant $C = C(a,b)$. Evidently,

$$\alpha = \alpha(a, b, p_1).$$

Note that, if the class $U(a,b)$is such that a is any real number, $b < -1$, then a stronger stability estimate may be proved (see Shishatskii and Atamanov, 1978).

§1.3 On the stability type for the Cauchy problem with data in the time half-axis.

The problem on the exactness of estimate (1.2.14) was solved by Shishatskii (1980). He had shown that the solution to the problem (1.0.1), (1.0.2) is really only logarithmically continuous by each $C^s[t \geq 0]$-norm of its Cauchy data even in the case when this solution together with its derivatives of every order is bounded in the quadrant $x \geq 0$, $t \geq 0$ and exponentially tends to zero for $t \to \infty$. For even n introduce the function

$$u(x,t) = \sum_{k=1}^{n} \frac{d_k}{n+2k} \exp\left\{\left[-1 + \frac{1}{(n+2k)^2+1}\right] t\right\} \sin(n+2k)x$$

$$+ \frac{\alpha_n + 1}{3n+1} \exp\left\{\left[-1 + \frac{1}{(3n+1)^2+1}\right] t\right\} \sin(3n+1)x, \qquad (1.3.1)$$

where for $k = 1, 2, ..., n$

$$\alpha_k = \prod_{1 \le i \le n_{i \ne k}} \left[\frac{1}{(n+2k)^2+1} - \frac{1}{(n+2i)^2+1} \right]^{-1}$$

$$\times \left[\frac{1}{(n+2k)^2+1} - \frac{1}{(3n+1)^2+1} \right]^{-1}, \qquad (1.3.2)$$

$$\alpha_{n+1} = \prod_{1 \le i \le n} \left[\frac{1}{(3n+1)^2+1} - \frac{1}{(n+2i)^2+1} \right]^{-1}.$$

It is easy to see that $u(x, t)$ solves equation (1.0.2) with the Cauchy data

$$f_1(t) = 0, \qquad (1.3.3)$$

$$f_2(t) = \sum_{k=1}^{n} \alpha_k \exp\left\{ \left[-1 + \frac{1}{(n+2k)^2+1} \right] t \right\}$$

$$+\alpha_{n+1} \exp\left\{ \left[-1 + \frac{1}{(3n+1)^2+1} \right] t \right\},$$

where $u(x, t)$ and its derivatives of each order are bounded for $x \ge 0$, $t \ge 0$ and for $t \to \infty$ decrease not more slowly than

$$O\left[\exp\left(-\frac{16}{17} \right) \right].$$

Further we shall need the estimates for $|\alpha_k|$ for $k = 1, 2, ..., n+1$. In order to obtain them we need the equalities

$$\left[1 + o\left(\frac{1}{n^2} \right) \right]^n = e^{0\left(\frac{1}{n}\right)},$$

$$a(a+1)(a+2)...b = e^{a-b} b^b a^{-a} e^{0(\ln a)},$$

$$a(a+2)(a+4)...b = e^{\frac{a}{2}-\frac{b}{2}} b^{\frac{b}{2}} a^{-\frac{a}{2}} e^{0(\ln a)},$$

For $k = 1, 2, ..., n$ we have

$$\alpha_k = \prod_{1\le i\le n_{i\neq k}} \left[\frac{1}{(n+2k)^2+1} - \frac{1}{(n+2i)^2+1}\right]^{-1}$$

$$\times \left[\frac{1}{(n+2k)^2+1} - \frac{1}{(3n+1)^2+1}\right]^{-1}$$

$$= \prod_{1\le i\le n_{i\neq k}} \frac{(n+2k)^2(n+2i)^2\left[1+(n+2k)^{-2}\right]}{(2n+2k+2i)|2k-2i|} \left[1+(n+2i)^{-2}\right] e^{0(\ln n)}$$

$$= \frac{(n+2k)^{2n}\left[(n+2)(n+4)...(3n)\right]^2}{(2n+2k+2)(2n+2k+4)...(4n+2k)2^n k!(n-k)!} e^{0(\ln n)}$$

$$= \frac{(n+2k)^{2n} e^{n-3n}(3n)^{3n}(2n+2k)^{n+k} e^{0(\ln 0)}}{n^n e^{n+k-2nk}(4n+2k)^{2n+k} 2^n k!(n-k)!}$$

$$< \frac{(n+2k)^{2n} e^{-n} 3^{3n} n^{3n} 2^{n+k}(n+k)^{n+k} e^{0(\ln n)}}{n^n 2^{(2n+k)}(2n+k)^{2n+k} 2^n e^{-n} n^n}$$

$$= 2^{-2n} 3^{3n} n^n \frac{(n+2k)^{2n}(n+k)^{n+k}}{(2n+k)^{2n+k}} e^{0(\ln n)}.$$

It may be shown that if k increases from 1 to n, and n is sufficiently large, then the fraction

$$(n+2k)^{2n}(n+k)(2n+k)^{(-2n+k)}$$

does not decrease. Therefore, for $k = 1, 2, ..., n$

$$|\alpha_k| < 3^{2n} n^{2n} e^{0(\ln n)}. \tag{1.3.4}$$

For α_{k+1} we obtain

$$\alpha_{n+1} = \prod_{1\le i\le n} \left[\frac{1}{(n+2i)^2+1} - \frac{1}{(3n+1)^2+1}\right]^{-1}$$

$$= \prod_{1\le i\le n} \frac{(n+2i)^2(3n+1)^2\left[1+(n+2i)^{-2}\right]}{(4n+2i+1)(2n-2i+1)} \left[1+(3n+1)^{-2}\right]$$

$$= \frac{(3n)^{2n}[(n+2)(n+4)...(3n)]^2}{(4n+2+1)(4n+4+1)...(6n+1).1.2...(2n-1)} e^{0(\frac{1}{n})}$$

$$= \frac{(3n)^{2n} e^{n-3n}(3n)^{3n}(4n)^{2n} e^{0(\ln n)}}{n^n e^{2n-3n}(6n)^{3n} e^{-n}(2n)^n}$$

$$= 3^{2n} n^{2n} e^{0(\ln n)}. \qquad (1.3.5)$$

Hence, in particular, it follows that inequality (1.3.4) holds for all $k = 1, 2, ..., n, n+1$.

We now introduce the notations

$$||f||_s = ||f(t)||_{C^s[t \geq 0]}, \qquad |||u(x,t)|||_m = ||u(x,t)||_{C^m[x \geq 0,\ t \geq 0]}$$

Inequality (1.3.4) helps to estimate $|||u(x,t)|||_m$ for every finite m. Really, evaluating in (1.3.1) the derivatives of the order $\leq m$ and replacing exponents and trigonometrical functions by unity, we obtain

$$|||u(x,t)|||_m < 3^{2n} n^{2n} e^{C_1 \ln n}, \qquad (1.3.6)$$

where C_1 depends on m. Noting that (n is even)

$$\left|\sin(n+2k)\frac{\pi}{2}\right| = 0, \qquad \left|\sin(3n+1)\frac{\pi}{2}\right| = 1,$$

and taking into account (1.3.5) we have

$$\left|u(\frac{\pi}{2}, 1)\right| > 3^{2n} n^{2n} e^{-C_1 \ln n}. \qquad (1.3.7)$$

Now estimate $C^s[t \geq 0]$ - norm of the Cauchy data for $u(x,t)$. By the first equality (1.3.3), it suffices to estimate the $f_2(t)$ norm. If

$$\hat{f}_2(p) = \int_0^\infty e^{-pt} f_2(t) dt,$$

then

$$\hat{f}_2(p) = \sum_{k=1}^{n} \frac{\alpha_k}{p+1-[(n+2k)^2+1]^{-1}} + \frac{\alpha_{n+1}}{p+1-[(3n+1)^2+1]^{-1}}$$

$$= \prod_{k=1}^{n} \left[p+1-\frac{1}{(n+2k)^2+1}\right]^{-1} \left[p+1-\frac{1}{(3n+1)^2+1}\right]^{-1}. \quad (1.3.8)$$

The coefficients $\alpha_1, ..., \alpha_n, \alpha_{n+1}$ in (1.3.2) are determined so that the latter equality holds. Note that for sufficiently large $|p|$

$$|\hat{f}_2(p)| < (|p|-1)^{-n-1}, \quad (1.3.9)$$

from which we conclude that $f_2(0) = 0$. This holds true because the integral

$$\frac{1}{2\pi i} \int_{a-i\infty}^{a+i\infty} e^{pt} \hat{f}_2(p) dp$$

for $a > 0$ uniformly converges and

$$f_2(0) = \lim_{t\to 0} f_2(t) = \frac{1}{2\pi i} \int_{a-i\infty}^{a+i\infty} e^{pt} f_2(p) dp.$$

Assuming $a \to \infty$ and taking into acoount (1.3.9) we see that the latter integral tends to zero. Consequently

$$\hat{f}_2'(p) = p\hat{f}_2(p).$$

Continuing this reasoning, we obtain that for $n \geq s$

$$\hat{f}_2^{(s)}(p) = p^s \hat{f}_2(p),$$

and, therefore, for $n \geq s+2$

$$|\hat{f}_2(ir)|, \quad |\hat{f}_2'(ir)|, ..., |\hat{f}_2^{(s)}(ir) < r^s(r+\frac{1}{2})^{-n}. \quad (1.3.10)$$

From (1.3.6) it follows that

$$|\hat{f}_2(iz)| < (r + \frac{1}{2})^{-n}.$$

Estimate (1.3.7) for $k = 0, 1, ..., s$ yields

$$|f_2^{(k)}(t)| = \left| \frac{1}{2\pi i} \int_{-i\infty}^{i\infty} e^{izt} \hat{f}_2^{(k)}(ir) dir \right| \le \frac{1}{2\pi} \int_{-\infty}^{\infty} r^s (r + \frac{1}{2})^{-n} dr$$

$$< \frac{1}{2\pi} \int_{-\infty}^{\infty} (r + \frac{1}{2})^{-n+s} dr < e^{C_3 n},$$

i.e., for $n \ge s + 2$

$$||u(0,t)||_s, \quad ||u_x(0,t)||_s < e^{C_3 n}. \tag{1.3.11}$$

Now let

$$v(x,t) = 3^{-2n} n^{-2n} e^{-C_1 n} u(x,t).$$

Then, (1.3.6), (1.3.7), (1.3.11) yield

$$|||v(x,t)|||_m < 1, \qquad \left| v(\frac{\pi}{2}, 1) \right| > e^{-C_4 \ln n},$$

$$||v(0,t)||_s, \qquad ||v_x(0,t)||_s < n^{-2n} e^{C_5 n}. \tag{1.3.12}$$

Let

$$n^{-2n} e^{C_5 n} = \varepsilon.$$

It is easy to verify that for sufficiently large n

$$\ln n < \ln \ln \frac{1}{\varepsilon}.$$

This and (1.3.12) yield the estimate

$$\left| v(\frac{\pi}{2}, 1) \right| > (\ln \frac{1}{\varepsilon})^{-C_4}.$$

The estimate (1.2.14) is not essentially improved.

§1.4 Existence of solutions in the class $U(a, b)$

In this section we shall give the proof of the existence of the solution to problem (1.0.1), (1.0.2).

Assume that in (1.0.1) $f_1(t) = 0$ and $f_2(t) = f(t)$. Then the following theorem holds.

Theorem 2.2.3. Let $f(t)$ be continuous together with its first derivative and

$$|f(t)|, \qquad |f'(t)| \leq Ce^{bt},$$

where $b < -1$. Then the function $u(x,t)$, defined by the formula

$$u(x,t) = -\int_t^\infty (f(\tau) + f'(\tau))e^{\tau - t} \int_0^x e^{2y-x} IJ dy d\tau, \tag{1.4.1}$$

where

$$I = I_0(\sqrt{2}(\tau - t)^{\frac{1}{2}} y^{\frac{1}{2}}); \quad J = J_0(\sqrt{2}(\tau - t)^{\frac{1}{2}}(x-y)^{\frac{1}{2}}),$$

solves problem (1.0.1), (1.0.2); and belongs to the class $U(1 + C\varepsilon^{-2}, b)$, where $\varepsilon^2 < -b - 1, C$ is a constant. Here $I_0(\xi)$ is the modified zero-order Bessel function; $J_0(\xi)$ is the zero-order Bessel function of the first genus.

Proof. First we show that $u(x,t)$ determined by (1.4.1) belongs to the class $U(1 + C\varepsilon^{-2}, b)$. To this end we have to show that

$$|u|, |u_\gamma| \leq Ce^{1 + C\varepsilon^{-2}x + bt}, \tag{1.4.2}$$

where $\gamma = t, x, xx, xt, xxt$.

Before we justify the differentiation in the integral (1.4.1), we obtain estimates for the Bessel function and its derivatives entering into (1.4.1). Since

$$J_0(x) = \frac{1}{2\pi} \int_{-\pi}^{\pi} e^{-ix\cos\psi} d\psi,$$

$$|J| = \left| J_0(\sqrt{2}(\tau - t)^{\frac{1}{2}}(x-y)^{\frac{1}{2}}) \right| \le 1. \tag{1.4.3}$$

Now evaluate I. As $J_0(ix) = I_0(x)$, then

$$I = I_0(\sqrt{2}(\tau - t)^{\frac{1}{2}} x^{\frac{1}{2}}) = J_0(i\sqrt{2}(\tau - t)^{\frac{1}{2}} x^{\frac{1}{2}})$$

$$= \frac{1}{2\pi} \int_{-\pi}^{\pi} \exp\left[\sqrt{2}(\tau - t)^{\frac{1}{2}} x^{\frac{1}{2}} \cos\psi\right] d\psi.$$

As a result we have

$$I \le e^{\sqrt{2}}(\tau - t)^{\frac{1}{2}} x^{\frac{1}{2}}. \tag{1.4.4}$$

It is well known that

$$I = \sum_{m=0}^{\infty} \frac{\left(\frac{\tau - t}{2}\right)^m x^m}{(m!)^2}. \tag{1.4.5}$$

Therefore,

$$I_t = -\frac{1}{2} x \sum_{m=1}^{\infty} \frac{m \left(\frac{\tau - t}{2}\right)^{m-1} x^{m-1}}{(m!)^2}. \tag{1.4.6}$$

Comparing series (1.4.5) and (1.4.6) we may say that series (1.4.5) majorizes the absolute values of series (1.4.6). Taking into account estimate (1.4.4), we have

$$|I_t| \le C x e^{\sqrt{2}(\tau - t)^{\frac{1}{2}} x^{\frac{1}{2}}}. \tag{1.4.7}$$

Since

$$J_t = -\frac{1}{2}(x - y) \sum_{m=1}^{\infty} \frac{m(-1)^m \left(\frac{\tau - t}{2}\right)^{m-1} (x-y)^{m-1}}{(m!)^2}, \tag{1.4.8}$$

then, comparing series (1.4.8) with (1.4.5) and taking into account (1.4.4), we obtain

$$|J_t| \leq C(x-y)e^{\sqrt{2}(\tau-t)^{\frac{1}{2}}(x-y)^{\frac{1}{2}}}. \tag{1.4.9}$$

For Jx we obtain the series of type (1.4.5)

$$J_x = \frac{\tau - t}{2}\sum_{m=1}^{\infty}\frac{m(-1)^m\left(\frac{\tau-t}{2}\right)^{m-1}(x-y)^{m-1}}{(m!)^2};$$

therefore,

$$|J_x| \leq C\frac{\tau-t}{2}e^{\sqrt{2}(\tau-t)^{1/2}(x-y)^{1/2}}. \tag{1.4.10}$$

Further,

$$J_{xx} = \frac{(\tau-t)^2}{4}\sum_{m=2}^{\infty}\frac{(-1)^m m(m-1)\left(\frac{\tau-t}{2}\right)^{m-2}(x-y)^{m-2}}{(m!)^2}$$

and, analogously,

$$|J_{xx}| \leq C(\tau-t)^2 e^{\sqrt{2}(\tau-t)^{\frac{1}{2}}(x-y)^{\frac{1}{2}}}. \tag{1.4.11}$$

Finally, for J_{xxt} we obtain the series

$$J_{xxt} = \frac{\tau-t}{4}\sum_{m=1}^{\infty}\frac{m^2(m-1)\left(\frac{\tau-t}{2}\right)^{m-2}(x-y)^{m-2}}{(m!)^2},$$

by means of which we have

$$|J_{xxt}| \leq C(\tau-t)\exp\sqrt{2}(\tau-t)^{\frac{1}{2}}(x-y)^{\frac{1}{2}}. \tag{1.4.12}$$

It is easy to show that

$$J_{xt} = -\frac{1}{2}J \qquad I_{xt} = -\frac{1}{2}I;$$

therefore,

$$|J_{xt}| = \leq \frac{1}{2}, \tag{1.4.13}$$

$$|J_{xt}| \le Ce^{\sqrt{2}(\tau-t)^{1/2}x^{1/2}}. \tag{1.4.14}$$

Now we shall justify the differentiation in (1.4.1). Introduce the notation

$$g(x,t,\tau) = (f(\tau)+f'(\tau))\, e^{\tau-t}\int\limits_0^x e^{2y-x} IJdy.$$

Then

$$u(x,t) = -\int\limits_t^\infty g(x,t,\tau)d\tau, \tag{1.4.15}$$

$$g_t(x,t,\tau) = (f(\tau)+f'(\tau))e^{\tau-t}$$

$$\times\left[-1+\int\limits_0^x e^{2y-x} I_t J dy + \int\limits_0^x e^{2y-x} IJ_t dy\right].$$

We now show that the integral in (1.4.15) converges in t. Taking into account (1.4.3) and (1.4.4) we obtain

$$|u(x,t)| = \left|\int\limits_t^\infty (f(\tau)+f'(\tau))\, e^{\tau-t}\int\limits_0^x e^{2y-x} IJdyd\tau\right|$$

$$\le \int\limits_t^\infty Ce^{b\tau}e^{\tau-t}\int\limits_0^x e^x e^{\sqrt{2}(\tau-t)^{\frac{1}{2}}x^{\frac{1}{2}}}dyd\tau$$

$$\le Cxe^{-t}e^x\int\limits_t^\infty e^{(b+1)\tau}e^{(\tau-t)\varepsilon^2+\frac{1}{2}\varepsilon^{-2}x}d\tau$$

$$= Cxe^{(1+\frac{1}{2}\varepsilon^{-2})x}e^{-(1+\varepsilon^2)t}\int\limits_0^\infty e^{(b+1+\varepsilon^2)\tau}d\tau$$

$$= Cx(b+1+\varepsilon^2)^{-1}e^{(1+\frac{1}{2}\varepsilon^{-2})x+bt} \le Ce^{(1+\varepsilon^2)x+bt}, \tag{1.4.16}$$

where $\varepsilon^2 < -1-b$.

Now we show that the integral

$$\int_t^\infty g_t(x,t,\tau)d\tau \tag{1.4.17}$$

uniformly converges in t. Using (1.4.5) and (1.4.8) we have

$$|g_t(x,t,\tau)| = |(f(\tau)+f'(\tau))|e^{\tau-t}$$

$$\times\left[1+\int_0^x e^{2y+x}I_tJdy+\int_0^x e^{2y-x}IJ_tdy\right]$$

$$\leq Ce^{bt}e^{\tau-t}\left(C\int_0^x e^x x e^{\sqrt{2}(\tau-t)^{1/2}x^{1/2}}dy + C\int_0^x xe^{2\sqrt{2}(\tau-t)^{1/2}x^{1/2}}dy\right)$$

$$= Ce^{b\tau}e^{\tau-t}(Cx^2e^xe^{\sqrt{2}(\tau-t)^{1/2}\varepsilon^{-1/2}x^{1/2}} + Cx^2e^x\exp\left[2\sqrt{2}(\tau-t)^{\frac{1}{2}}\frac{\varepsilon}{2}\frac{2}{\varepsilon}x^{\frac{1}{2}}\right]$$

$$\leq Ce^{b\tau}e^{\tau-t}(Cx^2e^xe^{\frac{x}{2\epsilon^2}}e^{(\tau-t)\varepsilon^2} + Cx^2e^x\exp\left[\frac{2x}{\varepsilon^2}+(\tau-t)\varepsilon^2\right]$$

$$\leq Ce^{bt}e^{\tau-t}e^{6x\varepsilon^{-2}}e^xe^{(\tau-t)\varepsilon^2} \leq Ce^{(b+1+\varepsilon^2)\tau+(1+6\varepsilon^{-2})x} = \varphi(\tau,\varepsilon).(1.4.18)$$

Since $b+1+\varepsilon^2<0$,the function $\varphi(\tau,x)$ is integrable in τ and is a majorant for $g_t(x,t,\tau)$. Therefore, integral (1.4.17) converges uniformly relative to t. Thus, the differentiability by t of the integral (1.4.1) is proved.

Consider the differentiation of (1.4.1) by x. From estimate (1.4.16) it follows that for $x\in[0,R]$, where R is an arbitrary but fixed number, the integral (1.4.15) converges uniformly relative to x.

Now we show the uniform convergence relative to x of the integral

$$\int_t^\infty g_x(x,t,\tau)d\tau. \tag{1.4.19}$$

Using (1.4.3), (1.4.4), (1.4.10) we have

$$|g_x(x,t,\tau)| = \left|(f(\tau)+f'(\tau))\, e^{\tau-t}\left(-\int_0^x e^{2y-x} IJdy\right.\right.$$

$$\left.\left.+\int_0^x e^{2y-x} IJdy\right) + e^x IJ(0)\right| \leq Ce^{b\tau}e^{\tau-t}\left(Cxe^x e^{\sqrt{2}(\tau-t)^{1/2}x^{1/2}}\right.$$

$$\left. + Cxe^x(\tau-t)e^{2\sqrt{2}(\tau-t)^{1/2}x^{1/2}} + Ce^x e^{\sqrt{2}(\tau-t)^{1/2}x^{1/2}}\right)$$

$$\leq Ce^{b\tau}e^{\tau-t}\left(Ce^{(\tau-t)\varepsilon^2+(1+\varepsilon^{-2})x} + Cxe^x e^{(\tau-t)\varepsilon^2+6\varepsilon^{-2}x}\right.$$

$$\left. + Ce^{(\tau-t)\varepsilon^2+(1+\frac{1}{2}\varepsilon^{-2})x}\right) \leq C\exp\left[(b+1+\varepsilon^2)\tau + (1+166\varepsilon^{-2})x\right]. \tag{1.4.20}$$

As $b+1+\varepsilon^2 < 0$, the right side of (1.4.20) is the integrable majorant of the function $g_x(x,t,\tau)$ for each $x \in [0,R]$, R being a fixed number. Therefore, (1.4.19) uniformly converges relative to x for $x \in [0,R]$ and for these x we may differentiate (1.4.1) under the integral sign. Because of the arbitrariness of R we obtain the result.

Analogously, we may verify that the integrals

$$\int_t^\infty g_{xx}(x,t,\tau)d\tau, \quad \int_t^\infty g_{xxt}(x,t,\tau)d\tau$$

uniformly converge in x and t.

We now need to show that (1.4.2) holds. Using estimate (1.4.8) we obtain

$$|u_t| = \left|\int_t^\infty g_t(x,t,\tau)d\tau\right| \leq \int_t^\infty Ce^{bt}e^{\tau-t}e^{(\tau-t)\varepsilon^2+(1+6\varepsilon^{-2})\varepsilon}d\tau$$

$$= Ce^{-(1+\varepsilon)^2 t}e^{(1+6\varepsilon^{-2})x}\int_t^\infty e^{(b+1+6\varepsilon^2)\tau}d\tau \leq Ce^{bt+(1+6\varepsilon^{-2})x}. \tag{1.4.21}$$

Further, using (1.4.17) we have

$$|u_x| = \left|\int_t^\infty g_x(x,t,\tau)d\tau\right| \le \int_t^\infty Ce^{bt}e^{\tau-t}e^{((\tau-t)\varepsilon^2+1+16\varepsilon^{-2})^x}d\tau$$

$$\le Ce^{-(1+\varepsilon)^2 t}e^{(1+16\varepsilon^{-2})x}\int_t^\infty e^{(b+1+16\varepsilon^{-2})\tau}d\tau \le Ce^{bt+(1+6\varepsilon^{-2})x}. \quad (1.4.22)$$

Analogously, taking into account (1.4.22), we have

$$|u_{xx}| \le ce^{bt+(1+36\varepsilon^{-2})x}.$$

Finally,

$$|u_{xxt}| \le c\exp[bt + (1+54\varepsilon^{-2})x].$$

An analogous estimate may also be made for $|u_{xt}|$. Thus, (1.4.2) holds true. Therefore, the function $u(x,t)$ determined by (1.4.1) belongs to the class $u(1+\varepsilon^{-2}, b)$.

Now by means of substitution, we shall show that $u(x,t)$ from (1.4.1) satisfies equation (1.0.2).

To simplify the calculation, transform formula (1.4.1). Since $J_0(ix) = I_0(x)$, then

$$J_0(\sqrt{2}(\tau-t)^{\frac{1}{2}}(x-y)^{\frac{1}{2}}) = I_0(\sqrt{2}(t-\tau)^{\frac{1}{2}}(x-y)^{\frac{1}{2}}) = \bar{I},$$

and, therefore, (1.4.1) may be rewritten in the form

$$u(x,t) = -\int_t^\infty (f(\tau)+f'(\tau))e^{\tau-t}\int_0^x e^{2y-x}I\bar{I}dyd\tau.$$

So, we have

$$u_t(x,t) = \int_t^\infty (f(\tau)+f'(\tau))e^{\tau-t}\int_0^x e^{2y-x}I\bar{I}dyd\tau$$

$$-\int_t^\infty (f(\tau)+f'(\tau))e^{\tau-t}\int_0^x e^{2y-x}I_t\bar{I}dyd\tau - \int_0^\infty (f(\tau)+f'(\tau))$$

$$\times \left[e^{\tau-t} \int_0^x e^{2y-x} I \bar{I}_t dy \right] d\tau + (f(t) + f'(t)) \int_0^x e^{2y-x} I(0)\bar{I}(0) dy;$$

$$u_x(x,t) = -\int_t^\infty (f(\tau) + f'(\tau)) e^{\tau-t}$$

$$\times \left[-\int_0^x e^{2y-x} I \bar{I} dy \int_0^x e^{2y-x} I \bar{I} dy + e^x I \bar{I}(0) \right] d\tau;$$

$$u_{xx} = -\int_t^\infty (f(\tau) + f'(\tau))$$

$$\times \Big[\int_0^x e^{2y-x} I \bar{I} dy - \int_0^x e^{2y-x} I I_x dy - e^x I \bar{I}(0) - \int_0^x e^{2y-x} I \bar{I}_x dy$$

$$+ \int_0^x e^{2y-x} I \bar{I}_{xx} dy + e^x I \bar{I}_x \Big|_{y=x} + e^x I \bar{I}_{(0)} + e^x I_x \bar{I}_0 \Big] d\tau.$$

Collecting similar terms and using the facts that

$$\bar{I}_x \big|_{y=x} = \frac{1}{2}(t-\tau), \qquad \bar{I}(0) = 1$$

we obtain

$$u_{xx} = -\int_t^\infty (f(\tau) + f'(\tau)) e^{\tau-t} \Big[\int_0^x e^{2y-x} I \bar{I} dy - 2 \int_0^x e^{2y-x} I \bar{I}_x d\dot{y}$$

$$+ \int_0^x e^{2y-x} I \bar{I}_{xx} dy + e^x I \exp \frac{1}{2}(t-\tau) + e^x I_x \Big] d\tau,$$

and

$$u_{xxt} = -u_{xx} - \int_t^\infty (f(\tau) + f'(\tau)) e^{\tau-t}$$

$$\times \Big[\int_0^x e^{2y-x} I_t \bar{I} dy + \int_0^x e^{2y-x} I \bar{I}_t dy - 2 \int_0^x e^{2y-x} I_t \bar{I}_x dy$$

$$-2 \int_0^x e^{2y-x} I \bar{I}_{xt} dy + e^x I_{xt} \Big] d\tau$$

$$+(f(t)+f'(t))\left[\int\limits_0^x e^{2y-x}I(0)\bar{I}(0)--2\int\limits_0^x e^{2y-x}I(0)\bar{I}_x(0)dy\right.$$

$$\left.+\int\limits_0^x e^{2y-x}I(0)\bar{I}_{xx}(0)dy+e^xI(0)+e^xI_x(0)\right].$$

Substituting the obtained estimates in the equation (1.0.2); taking into account that $\bar{I}(0)=\bar{I}_x(0)=0$; and collecting the similar terms, we can show that all we need to prove is that the following expression

$$-\int\limits_t^\infty (f(\tau)-f'(\tau))e^{\tau-t}\left[\int\limits_0^x e^{2y-x}I\bar{I}dy-2\int\limits_0^x e^{2y-x}I_t\bar{I}_xdy\right.$$

$$-2\int\limits_0^x e^{2y-x}I\bar{I}_{xt}dy-\int\limits_0^x e^{2y-x}I_t\bar{I}_{xx}dy$$

$$\left.+\int\limits_0^x e^{2y-x}I\bar{I}_{xxt}+\frac{1}{2}e^xI_t(t-\tau)+\frac{1}{2}e^xI+e^xI_{xt}\right]$$

is equal to zero.

If we use the relations

$$I_0'(\xi)=I_1(\xi),\quad I_2(\xi)=I_0(\xi)=2\xi^{-1}I(\xi),$$

$$I_1'(\xi)=\frac{1}{2}I_0+\frac{1}{2}I_2,\quad I_{xt}=-\frac{1}{2}I,\qquad \bar{I}_{xt}=\frac{1}{2}\bar{I};$$

then the expression takes the form

$$-\int\limits_t^\infty (f(\tau)+f'(\tau))e^{\tau-t}\left[-2\int\limits_0^x e^{2y-x}I_tI_xdy+\int\limits_0^x e^{2y-x}I_t\bar{I}_{xx}dy\right.$$

$$\left.+\int\limits_0^x e^{2y-x}I\bar{I}_{xxt}dy+\frac{1}{2}e^xI_t(t-\tau)\right].$$

Further, we use that

$$\bar{I'}_x = -I'_y, \quad I_{xy}\bar{I}_x = -\frac{1}{2\sqrt{2}} I I_1 (t-\tau)^{\frac{1}{2}}(x-y)^{-\frac{1}{2}},$$

$$\bar{I}_{xxt} = \frac{1}{2\sqrt{2}} \bar{I}_1 (t-\tau)^{\frac{1}{2}}(x-y)^{-\frac{1}{2}}.$$

Integrating by parts we obtain

$$-\int_t^\infty (f(\tau)+f'(\tau))e^{\tau-t}\Big[-2\int_0^x e^{2y-x} I_t \bar{I}_x dy - \int_0^x e^{2y-x} I_t \bar{I}_{xy} dy$$

$$+\int_0^x e^{2y-x}\bar{I}_{xxt} dy e^x I_t \frac{t-\tau}{2}\Big] d\tau = -\int_t^\infty (f(\tau)+f'(\tau))e^{\tau-t}$$

$$\times\Big[-2\int_0^x e^{2y-x} I_t \bar{I}_x dy - e^{2y-x} I_t \bar{I}_x\Big|_0^x + 2\int_0^x e^{2y-x} I_t \bar{I}_x dy$$

$$+\int_0^x e^{2y-x} I_{ty}\bar{I}_x dy + \int_0^x e^{2y-x}\bar{I}_{xxt} dy + e^x I_t \frac{t-\tau}{2}\Big] d\tau$$

$$= -\int_y^\infty (f(\tau)+f'(\tau))e^{\tau-t}\Big[-\frac{1}{2}\int_0^x e^{2y-x}\Big(\frac{t-\tau}{2}\Big)^{1/2}(x-y)^{1/2} dy$$

$$+\frac{1}{2}\int_0^x e^{2y-x} I\bar{I}_1\left(\frac{t-\tau}{2}\right)\frac{1}{2}(x-y)^{1/2} dy$$

$$-e^x I_t \frac{t-\tau}{2} + +e^x I_x \frac{t-\tau}{2}\Big] d\tau = 0.$$

So, the function $u(x,t)$, determined by (1.4.1) satisfies the equation.

Now we verify the conditions (1.0.1). The first condition follows from substitution of $x=0$ into (1.0.2) and the fact that $u(x,t)\to\infty$ uniformly for $x\to 0$.

$$u_x(0,t) = -\int_t^\infty (f(\tau)+f'(\tau))e^{\tau-t} I(0)\bar{I}(0) dt$$

$$= e^{-t}\Big[\int_t^\infty f(\tau)e^\tau d\tau + \int_t^\infty f'(\tau)e^\tau d\tau\Big]$$

$$= e^{-\tau}\Big[\int_t^\infty f(\tau)e^\tau d\tau + f(\tau)e^\tau\Big|_t^\infty - \int_t^\infty f(\tau)e^\tau d\tau\Big]$$

$$= e^{-t}(-e^t f(t)) = f(t).$$

By means of the obtained estimates we see that u_x uniformly converges to $f(t)$ for $x \to 0$.

§1.5 The Cauchy problem in the multidimensional case.

Consider the two-dimensional pseudoparabolic equation

$$(I - \Delta)u_t(x, y, t) = \Delta u(x, y, t) \tag{1.5.1}$$

in the domain

$$D = \{(x, y, t) : \ 0 \le x < \infty, \quad -\infty < y < \infty, \quad t \ge 0\}$$

with the data

$$u(0, y, t) = f_1(y, t), \quad u_x(0, y, t) = f_2(y, t). \tag{1.5.2}$$

We shall reduce problem (1.5.1), (1.5.2) by means of the integral transformation to a one-dimensional one with the parameter and use the obtained above results.

Definition. We say that $u(x, y, t) \in U_1(a, b)$ if u_1, $u\gamma$ are continuous in D and

$$|u|, |u_\gamma| \le C_u e^{ax+bt}(t + y^2)^{-1}, \tag{1.5.3}$$

where $\gamma = x, y, t, xx, yy, xt, yt, xxt, yyt$. The numbers a, b satisfy the following condition: if $a < 1$, then b is any real number; if $a > 1$, then $b < -1 + [4a(a + \sqrt{a^2 - 1}]^{-1}$. C_u is a constant depending on u.

Introduce a notation

$$\hat{u}(x, \xi, t) = \frac{1}{\sqrt{\pi}} \int_{-\infty}^{\infty} e^{-i\xi y} u(x, y, t) dy, \ -\infty < \xi < \infty. \tag{1.5.4}$$

Applying to (1.5.1), (1.5.2) the transformation by y, we obtain the following problem

$$\hat{u}_t(x, \xi, t) = \hat{u}_{xxt}(x, \xi, t) - \xi^2 \hat{u}_t(x, \xi, t) + +\hat{u}_{xx}(x, \xi, t) - \xi^2 \hat{u}(x, \xi, t), \tag{1.5.5}$$

$$\hat{u}(0, \xi, t) = \hat{f}_1(\xi, t), \quad \hat{u}_x(0, \xi, t) = \hat{f}_2(\xi, t). \tag{1.5.6}$$

Introduce the new function $V(x, \xi, t)$ defined by the formula

$$\hat{u}(x, \xi, t) = V(x, \xi, t) \exp(-\frac{\xi^2 t}{1 + \xi^2}).$$

Then, in terms of this function, (1.5.5), (1.5.6) will have the following form

$$(1 + \xi^2) V_t(x, \xi, t) = V_{xxt}(x, \xi, t) + (1 + \xi^2)^{-1} V_{xx}(x, \xi, t),$$

$$V(0, \xi, t) = \hat{f}_1 \exp[(1 + \xi^2)^{-1} \xi^2 t],$$

$$V_x(0, \xi, t) = \hat{f}_2 \exp[(1 + \xi^2)^{-1} \xi^2 t].$$

Now, change variables using

$$z = (1 + \xi^2)^{\frac{1}{2}} x, \quad \tau = (1 + \xi^2)^{-1} t.$$

As a result we obtain

$$V_\tau(z,\xi,\tau) = V_{zzz}(z,\xi,\tau) + V_{zz}(z,\xi,\tau), \tag{1.5.7}$$

$$V(0,\xi,\tau) = F_1(\xi,\tau), \quad V_z(0,\xi,\tau) = F_2(\xi,\tau). \tag{1.5.8}$$

Condition (1.5.3) for the function $V(z,\xi,\tau)$ will be as follows

$$|V|,|V_\tau|,|V_z|,|V_{z\tau}|,|V_{zz}|,|V_{zzz}| \leq C\exp\Big[\frac{az}{\sqrt{1+\xi^2}}$$

$$+b+\xi^2(1+\xi^2)^{-1}(1+\xi^2)\tau\Big] = C\exp\Big[\frac{az}{\sqrt{1+\xi^2}}\big[b+(b+1)\xi^2\big]\tau\Big]. \tag{1.5.9}$$

Let $u(x,y,z) \in U_1(a,b)$. We must show that, in this case, $V \in U(a,b)$. If $a<1$, then $a(1+\xi^2)^{-1/2} < 1$ for each ξ and $b+(b+1)\xi^2$ may attain any value. Assume $a \geq 1$; then, for ξ such that $\xi^2 > a^2-1$, we have $a(1+\xi^2)^{-\frac{1}{2}} < 1$ and, therefore, for each value of $b+(b+1)\xi^2$ the function $V(z,\xi,\tau)$ will belong to the class $U(a,b)$. Now, let $a \geq 1$ and $a(1+\xi^2)^{-\frac{1}{2}} \geq 1$. We must show that, in this case, from

$$b < -1 + [4a(a+\sqrt{a^2-1})]^{-1} \tag{1.5.10}$$

will follow

$$b+(b+1)\xi^2 < -1 + \Big[\frac{4a}{\sqrt{1+\xi^2}}\Big(\frac{a}{\sqrt{1+\xi^2}}+$$

$$+\big(a^2(1+\xi^2)^{-1}-1\big)^{-\frac{1}{2}}\Big)\Big]^{-1}. \tag{1.5.11}$$

Really, using (1.5.10), we obtain

$$b+(b+1)\xi^2 < -1 + [4a(a+\sqrt{a^2-1})]^{-1}$$

$$+(1+[4a(a+\sqrt{a^2-1})]^{-1}+1)\xi^2$$

$$= -1 + [4a(a+\sqrt{a^2-1})]^{-1} + \xi^2[4a(a+\sqrt{a^2-1})]^{-1}$$

$$< -1 + [4a(a + \sqrt{a^2 - 1 - \xi^2})]^{-1} + \xi^2[4a(a + \sqrt{a^2 - 1 - \xi^2})]^{-1}$$

$$= -1 + \sqrt{1 + \xi^2}[4a(a(1 + \xi^2)^{-\frac{1}{2}} + a^2(1 + \xi^2)^{-1} - 1)^{\frac{1}{2}}]^{-1}$$

$$= -1 + \left[\frac{4a}{\sqrt{1+\xi^2}}\left(\frac{a}{\sqrt{1+\xi^2}} + \sqrt{\frac{a^2}{1+\xi^2} - 1}\right)\right]^{-1}.$$

Thus, relation (1.5.11) is established. So, if $u \in U_1(a, b)$, then $V \in U(a, b)$. So, we may reformulate the results from the previous sections for the two-dimensional case.

Theorem 1.5.1. The solution to the problem (1.5.1), (1.5.2) is unique in the class of functions $U_1(a, b)$.

For the solutions from the fixed class $U_1(a, b)$ and such that $C_u \leq C$, where C is an absolute constant, the following theorem holds.

Theorem 1.5.2. For the solution $u(x, y, t) \in U_1(a, b)$ to the problem (1.5.1), (1.5.2) let the following conditions

$$C_u = 1, |f_1|, |f_2|, |f_1'|, |f_2'| \leq \varepsilon e^{bt}(1 + y^2)^{-1}, \quad 0 \leq t \leq \infty,$$

hold (ε is sufficiently small). Then, such $p_1 > 0$, $C = C(a, b)$, $\alpha = \alpha(a, b) > 0$ exist that

$$|u(x, y, t)| \leq C \left(\frac{\ln \ln \varepsilon^{-\alpha}}{\ln \varepsilon^{-\alpha}}\right)^{\frac{1}{2}} e^{p_1 x + bt}.$$

Now consider the three-dimensional pseudoparabolic equation

$$(I - \Delta)u_t(x, y, z, t) = \Delta u(x, y, z, t) \tag{1.5.12}$$

in the domain

$$D = \left\{(x, y, z, t), \quad 0 \leq x < \infty, -\infty < y < \infty, -\infty < z < \infty, \quad t \geq 0\right\}$$

with the data in the time-like manifold

$$u(0, y, z, t) = f_1(y, z, t), \; u_x(0, y, z, t) = f_2(y, z, t). \tag{1.5.13}$$

Definition. We say that $u(x,y,z,t) \in U_2(a,b)$ if u, u_γ, are continuous in D and

$$|u|,\ |u_\gamma| \leq C(1+y^2)^{-1}(1+z^2)^{-1}\exp(ax+bt),$$

where a,b satisfy the following conditions: if $a<1$, then b is any real number; if $a \geq 1$, then

$$b < -1 + [4a(a-\sqrt{a^2-1})]^{-1}, \quad \gamma = x,y,z,t,xx,yy,zz,zzt,xxt,yyt.$$

It may be shown that in the class of functions $U_2(a,b)$ for the problem (1.5.12), (1.5.13) the analogous theorems of uniqueness and conditional stability hold.

§1.6 Uniqueness of continuation of solution to the degenerate differential equations with unbounded operator in the characteristic case.

The results in this section belong entirely to S.P.Shishatskii and he had intended to publish them.

Let H be a real Hilbert space with a scalar product $(.,.)$ and a norm $\|\ \|$. We assume that D is a dense linear set from H and $[.,.]$ is a non-negative bilinear form in D (D may not be closed relative to this form). Let $A(t)$ be a linear operator symmetric in D, $A(t): D \to H$, and $A(t)$ satisfy the following conditions

$$(-A(t)h,h) \geq a_1[h,h] - a_2\|h\|^2, \quad (h \in D, \ \ a_1 > 0, \ \ a_2 \geq 0), \tag{1.6.1}$$

$$|(A_t(t)h,h)| \leq a_3 t^{-1}\ln^{-2}\frac{1}{t}[h,h], \ \ (h \in D, \ \ a_3 \geq 0). \tag{1.6.2}$$

The derivative $A_t(t)$ is considered as the corresponding limit in D in the norm from H. We assume that this limit exists.

Consider the problem of uniqueness of solution $u(t)$ continuation of the inequality

$$\|tu_t + A(t)u\|^2 \leq \ln^{-2}\frac{1}{t}(b_1[u,u] + b_2\|u\|^2) \tag{1.6.3}$$

from the point $t = 0$. The simplest example of such inequality is the equation

$$tu_t + u_{xx} = 0,$$

taken for functions that are periodic in x. The solution to the Cauchy problem with the data for $t = 0$ for this equation is not unique. For each integer n we have the periodic in x solution

$$u(t,x) = t^{n^2}\sin nx$$

with the zero Cauchy data. It turns out that in this case a theorem similar to the Aronzine-Cordes theorem for elliptic equations holds.

Theorem 1.6.1. Let $u(t)$ solve inequality (1.6.3) such that for each $\lambda > 0$

$$t^{-\lambda}u(t) \to 0, \quad (t \to 0).$$

Then $u(t) \equiv 0$.

Proof. We begin by establishing the inequality

$$\gamma e^{2x}\|\alpha u_t + Au\{[(\alpha\gamma)_t + 2\gamma\Psi] - (Au,u)$$

$$-\alpha\gamma(A_t u,u) + [(\alpha^2\gamma x_t)_t + 2\alpha\gamma x_t\Psi - \gamma\Psi^2]\|u\|^2\}$$

$$+\{\alpha\gamma e^{2x}[(Au,u) - \alpha x_t\|u\|^2]\}_t, \tag{1.6.4}$$

where $\alpha(t), x(t), \gamma(t), \Psi(t)$ are smooth scalar functions. Setting $V = e^x u$ we obtain

$$\gamma e^{2x}\|\alpha u_t + Au\|^2 = \gamma\|\alpha(V_t + x_t V + AV)\|^2$$

$$= \gamma[\|AV - \alpha x_t V + \Psi V - \Psi V\|^2 + \alpha^2\|V_t\|^2 + 2\alpha(AV - \alpha x_t V, V_t)]$$

$$\geq \gamma[-2\Psi(AV, V) + 2\alpha(AV, V_t) - 2\alpha^2 x_t(V, V_t)$$

$$-\Psi^2\|V\|^2 + 2\alpha\Psi x_t\|V\|^2].$$

Using the equalities

$$2\alpha\gamma(AV, V_t) = -(\alpha\gamma)_t(AV, V) - \alpha\gamma(A_t V, V) + [\alpha\gamma(AV, V)]_t,$$

$$2\alpha^2\gamma x_t(V, V_t) = (\alpha^2\gamma x_t)_t\|V\|^2 - \alpha^2\gamma x_t\|V\|^2)_t.$$

we obtain (1.6.4). Setting in this inequality

$$\alpha = t,\ x = -\lambda\left[\ln t + 2(k+1)\ln\ln\frac{1}{t}\right],$$

$$\gamma = t^{-1},\quad \Psi = k\ln^{-2}\frac{1}{t},\quad (\lambda > 0,\ k \geq 0),$$

evaluating its right-side coefficients, and taking into account (1.6.2) and (1.6.3), we obtain the estimate

$$t^{2\lambda-1}\ln^{-2(k+1)\lambda-2}\frac{1}{t}\Big\{2ka_1 - a_3 - b_1)[u,u] + 2\lambda(1 - 2ka_2\lambda^{-1}$$

$$+b_2\lambda^{-1} - \frac{1}{2}k^2\ln^{-2}\frac{1}{2})\|u\|^2\Big\} \leq \Big\{t^{2\lambda}\ln^{-2(k+1)\lambda}\frac{1}{t}$$

$$\times\left[(-Au, u) + \lambda(-1 + 2(k+1)\ln^{-1}\frac{1}{t}\|u\|^2\right]\Big\}_t.$$

Setting

$$k = (a_3 + b_1)2a_1^{-1}$$

and taking λ sufficiently large and t sufficiently small, we obtain

$$\lambda t^{-2\lambda-1} \ln^{-2(k+1)\lambda-2} \frac{1}{t} \|u\|^2$$

$$\leq \left\{ t^{2\lambda} \ln^{-2(k+1)\lambda} \frac{1}{t} \left[(-Au, u) + \lambda(-1 + 2(k+1) \ln^{-1} \frac{1}{t} \|u\|^2 \right] \right\}_t .$$

Conventional arguments now give $u(t) \equiv 0$.

Remark. If in (1.6.3) $b_1 = 0$, then the factor $b_2 \ln^{-2} \frac{1}{t}$ may be replaced by a constant C and the result of the theorem still holds, but with certain additional condition on $A(t)$. Namely, let $A(t) \equiv A$ be not dependent on t, self-conjugated (σ is the spectrum) and

$$\sigma(-A) \subset (-\infty, \infty) - \bigcup_1^{\infty} (C_n, d_n).$$

The intervals (C_n, d_n) are such that

$$C_{n+1} - d_n = 2h_n \geq 2h_0 > 0.$$

Then the following simple estimate holds

$$\lambda_n = 0.5(C_n + d_n),$$

$$t^{-1-2\lambda_n} \|tu_t + Au\|^2 \geq h_n^2 t^{-1-2\lambda_n} \|u\|^2$$

$$+ \left\{ t^{-2\lambda_n} [(Au, u) + \lambda_n \|u\|^2] \right\}_t . \tag{1.6.5}$$

Really, ($V = t^{-\lambda_n} u$)

$$t^{-1-2\lambda_n}\|tu_t - Au\|^2 = t^{-1}\|tV_t + AV + \lambda_n V\|^2$$

$$= t\|V\|^2 + 2(V_t, AV = \lambda_n V) + t^{-1}\|AV + \lambda_n V\|^2$$

$$\geq t^{-1-2\lambda_n}\|Au + \lambda u\|^2 + \left\{t^{-2\lambda_n}\left[(Au, u) + \lambda_n\|u\|^2\right]\right\}_t$$

$$\geq h_n^2 t^{-1-2\lambda}\|u\|^2 + \left\{t^{-2\lambda_n}\left[(Au, u) + \lambda_n\|u\|^2\right]\right\}_t,$$

and the remark follows from estimate (1.6.5), analogous to Theorem 1.6.1.

Theorem 1.6.2. Let $p > 1$ and $u(t)$ be the solution to the inequality

$$\|t^p u_t + Au\|^2 \leq t^{p-1}\ln^{-2}\frac{1}{t}(b_1[u, u] + b_2\|u\|^2)$$

such that for each $\lambda > 0$

$$\exp(\lambda t^{-p+1})u(t) \to 0, \quad (t \to 0).$$

Then $u(t) \equiv 0$.

The proof follows from estimate (1.6.4), where we set

$$\alpha = t^p,\ x = -\lambda t^{-p+1}\left[1 + (2k+1)\ln^{-1}\frac{1}{t}\right],$$

$$\gamma = t^{-p},\quad \Psi = kt^{p-1}\ln^{-2}\frac{1}{t},\quad (\lambda > 0,\ k \geq 0).$$

Consider now the second-order equations (conditions (1.6.1) and (1.6.2) we assume to hold).

Theorem 1.6.3. Let the solution $u(t)$ of the inequality

$$\|t^2 u_{tt} + Au\|^2 \leq \ln^{-2}\frac{1}{t}(b_1 t^2\|u_t\|^2 + b_2[u,u] + b_3\|u\|^2) \qquad (1.6.6)$$

be such that for each $\lambda > 0$

$$t^{-\lambda}u(x,t), \quad t^{-\lambda}u_t(x,t) \to 0 \quad (t \to 0).$$

Then $u(t) \equiv 0$.

Proof of this theorem is based on the following estimate

$$\gamma e^{2x}\|\alpha u_{tt} + Au\|^2 \geq 2e^{2x}\Big\{\Big[(\alpha^{2\gamma})_t x_t$$

$$+\alpha\gamma(\alpha x_{tt} + \Psi)\Big]\|u_t\|^2 + \Big[-(\alpha\gamma)_t x_t$$

$$+\gamma(-\alpha x_{tt} + \Psi)\Big](-Au, u) + \alpha\gamma x_t(A_t u, u)$$

$$+\Big[2\alpha\gamma x_t^2(\alpha x_{tt} - \Psi) - t^{-1}\Big[t(\alpha^2\gamma x_t^2)_t\Big]_t + \frac{1}{2}\alpha^2\gamma t^{-2}x_t^2$$

$$-(\alpha\gamma\Psi)_t x_t - \frac{1}{2}\gamma\Psi^2\frac{1}{2}(\alpha\gamma\Psi)_{tt})\Big]\|u\|^2\Big\} + F_t(t). \qquad (1.6.7)$$

The function $e^{-x}F(t)$ here is the square form relative to the vector (u_t, Au, u). Its exact form in our case is not essential. The method for proof of estimate (1.6.7) is analogous to those for estimate (1.6.4). To prove Theorem 1.6.3 in (1.6.7) we have to set

$$\alpha = t^2, \quad x = -\lambda(\ln t + 2x\ln\ln\frac{1}{t}), \quad \gamma = t^2\exp(2k\ln^{-1}\frac{1}{t}),$$

$$\Psi = \lambda(1 - 2k\ln^{-1}\frac{1}{t} + k\ln^{-2}\frac{1}{t}), \quad (\lambda > 0, \ k \geq 0).$$

Remark. If in (1.6.6) we set $b_1 = b_2 = 0$,then the coefficient $b_3\ln^{-2}\frac{1}{t}$ may be replaced by the constant and Theorem 1.6.3 will be valid. The proof of this fact is analogous to those in Remark to Theorem 1.6.1.

Theorem 1.6.3. gives the following strengthening of the Aronzine-Cordes theorem.

Theorem 1.6.4. Let $u(x) \in C^2$ be a solution to the inequality

$$\left|\sum \frac{\delta}{\delta x_i} a^{ij} \frac{\delta u}{\delta x_j}\right|^2 \leq \ln^{-2} \frac{1}{|x|}(b_1|x|^{-2}|\nabla u|^2 + b_2|x|^{-4}u^2)$$

where $a^{ij} \in C^2$, the matrix (a^{ij}) be positively determined; b_1, b_2 be arbitary. If, for each $\lambda > 0$

$$|x|^{-\lambda}u(x), \quad |x|^{-\lambda}\nabla u(x) \to 0 \quad (|x| \to 0),$$

then $u(x) \equiv 0$.

Theorem 1.6.5. Let $p > 2$ and the solution $u(t)$ to the inequality

$$|t^p u_{tt} + Au|^2 \leq t^{\frac{p}{2}-1} \ln^{-2} \frac{1}{t}(b_1 t^p \|u_t\|^2 + b_2[u,u] + b_3\|u\|^2)$$

be such that for each $\lambda > 0$

$$\exp(\lambda t^{-p/2+1})\{u(t), u_t(t)\} \to 0, \quad (t \to 0).$$

Then $u(t) \equiv 0$.

To prove this theorem we again estimate (1.6.7), where we set

$$\alpha = t^p,\ x = \lambda t^{-\frac{p}{2}+1}(1 - k\ln^{-1}\frac{1}{t}),$$

$$\gamma = t^{-p}\exp(k\ln^{-1}\frac{1}{t}), \quad (\lambda > 0,\ k \geq 0).$$

All the above results were formulated as theorems of operator-differential equation theory. This is because parallel with the required decreasing of solution for $t \to 0$ we have to satisfy some conditions relative to transversal variables (periodicity, I boundary-value problem conditions and so on). The following problem is now of interest: how can we weaken these conditions?

We shall give here two answers to this question. The first of them may be considered as a type of Fragmen-Lindelef theorem for elliptic equations. It is required there that the domain, where the solution decreases in t, have to be sufficiently large with respect to x.

The second, which relates to parabolic equations, has no additional conditions for variable x.

Theorem 1.6.6. Let a solution $u(t,x) \in C^2$ to the inequality

$$(t^2 u_{tt} + u_{xx})^2 \le bt^{-\omega} u^2$$

be such that for $-\pi(2\omega)^{-1} \le x \le \pi(2\omega)^{-1}$ the following inequalities

$$u_t^2(t,x),\ \ u_x^2(t,x),\ \ u^2(t,x) \le C\exp(-t^{-\Omega}),\ \ (\Omega > \omega).$$

hold. Then $u(t,x) \equiv 0$.

The proof of the theorem is based on the estimate

$$t^{-2+2\omega}\exp(2\lambda t^{-\omega}\cos\omega x)(t^2 u_{tt} + uxx)^2$$

$$\ge \omega^2 \lambda t^{-2+\omega}\cos\omega x \exp(2\lambda t^{-\omega}\cos\omega x)^2 u + \div u.$$

Here both component of the vector $\exp(-2\lambda t^{-w}\cos\omega x)u$ are the square forms relative to the vector (u_t, u_x, u).

Theorem 1.6.7. Let $p \ge 2$, $\Omega > p-2$, $\Theta < \frac{1}{2}\Omega$. If $u(x,t) \in C^2$ solves the inequality

$$(t^p u_{tt} + u_x)^2 \le b_1 t^{2p-\Theta-2} u_t^2 + b_2 t^{2p-3\Theta-4} u^2,$$

and

$$u_t^2(t,x),\ u_x^2(t,x),\ u^2(t,x) < C\exp(-t^{-\Omega}),$$

then $u(t,x) \equiv 0$.

The proof of this theorem uses the following estimate

$$t^{-p}\exp(2\lambda t^{-w}-2\lambda^2x^2)(t^p u_{tt}+u_x)^2$$

$$\geq 2\omega\lambda\exp(2\lambda t^{-\omega}-2\lambda^2x^2)\left\{(\omega-\frac{p-2}{2})t^{p-\omega-2}u_t^2\right.$$

$$\left.+2\omega^2(\omega-\frac{p-2}{2})\lambda^2 t^{p-3\omega-4}[1+0(1)]u^2\right\}+\div u$$

Here U is the vector analogous to those in the estimate of Theorem 1.6.6.

Chapter 2

Interior problems for pseudoparabolic equations.

We shall denote, as interior problems for differential equations, problems of the following type (see Lavrent'ev and Amonov, 1982). Let there be given

1) a differential equation;
2) a domain D where a certain solution to this equation exists;
3) a set A lying inside the set $D(A \subset D)$;
4) the values of solutions in the set A, i.e.,

$$u\Big|_A = f,$$

where f is known. The problem is to find the solution in the whole D.

Many applications lead to the interior problems. First, may be interpretation of the physical observation data. Often it occurs that the physical field satisfies a certain equation. The field is measured only in some set - for example, in a set of lines. It is required to find the field in the whole domain. Such situation arises when using aerocosmic information, when the data from a satellite or aeroplane come from the curve, the trajectory of transmitter support.

The classical interior problems for differential equations are the problems of analytic continuation of holomorphic functions of one complex variable. The theory of these problems is now rather complete.

The classical interior uniqueness theorem gives the necessary and sufficient condition for the set, which belongs to the domain of regularity, from which the analytic continuation of a holomorphic function is unique. The solution stability of this problem was considered by Lavrent'ev (1962) and others.

The theory of interior problems for differential equations with partial derivatives relative to many independent variables is intensively developed nowdays. Some theorems of uniqueness and stability for interior problems for elliptic equations were obtained by Lavrent'ev and Lavrent'ev (1976), Abdukarimov (1983), Shishatskii (1977), and others. Interior problems for parabolic equations were considered by Lavrent'ev and Amonov (1976, 1982); Shishatskii and others. The problems of continuation of solutions to hyperbolic equations have been investigated by Ptashnic (1984).

§2.1 Three-point problem.

In this section we shall be interested in the uniqueness and the conditional stability estimate for the following three-point problem. Let $u(x,t)$ be given in the semi-strip

$$D = \{0 \leq x \leq \pi,\ 0 \leq t < \infty\}$$

and let it be a solution to equation (1.0.1) with the following conditions

$$u(0,t) = u(\pi,t) = 0, \tag{2.1.1}$$

$$u(x^*,t) = f(t), \tag{2.1.2}$$

Here x^* is a given number from $[0,\pi]$; $f(t)$ is a given function. Our problem will be to find the function $u(x,t)$ in the whole D, given (2.1.1) and (2.1.2). As it is easy to see, the solution to problem (1.0.1), (2.1.1), (2.1.2) will be unique (or not) if x^*/π will be irrational (rational). It turns out to be that if x^*/π is not only irrational, but, additionally, is badly approximated by rational numbers, then the problem additionally has the stability estimate of the Holder type by the variations of the function $f(t)$.

Definition. We say that a function $u(x,t)$ belongs to the class $u(M)$ if u, u_γ are continuous and

$$|u|, |u_\gamma| \leq M$$

in D, where $\gamma = x, t, xx, xt, xxt, xxx$.

Theorem 2.1.1. The solution to problem (1.0.1), (2.1.1), (2.1.2) for irrational x^*/π is unique in the class $u(M)$.

Proof. We take the solution to equation (1.0.1) with conditions (2.1.1) from the class $u(M)$ in the form

$$u(x,t) = \sum_{k=1}^{\infty} a_k e^{-\frac{k^2 t}{k^2+1}} \sin kx, \tag{2.1.3}$$

where a_k are the Fourier coefficients of $u(x,0)$. Formulae (2.1.2), (2.1.3) yield

$$f(t) = \sum_{k=1}^{\infty} a_k e^{-\frac{k^2 t}{k^2+1}} \sin kx^*. \tag{2.1.4}$$

Evidently, the solution will not be unique if x^*/π be rational. So, we shall consider x^*/π as irrational. Apply to (2.1.4) the Laplace trasformation

$$\sum_{k=1}^{\infty} a_k \sin kx^* \int_0^{\infty} e^{-(p+\frac{k^2}{k^2+1})t} dt = F(p), \tag{2.1.5}$$

where $F(p)$ is the image of the function $f(t)$ after the Laplace transformation. Integrate both sides of (2.1.5) with respect to the circles

$$\Gamma_n = \left\{p : |p - p_n| = \frac{1}{2}\left[\frac{(n+1)^2}{(n+1)^2+1} - \frac{n^2}{n^2+1}\right]\right\}, \tag{2.1.6}$$

with centers in the points $p_n = -n^2(1+n^2)^{-1}$. As a result, we obtain

$$\sum_{k=1}^{\infty} a_k \sin kx^* \int_{\Gamma_n} [p + k^2(k^2+1)^{-1}]^{-1} dp = \int_{\Gamma_n} F(p)dp.$$

This equation and the integral Cauchy formula yield

$$2\pi i a_n \sin nx^* = \int_{\Gamma_n} F(p)dp. \tag{2.1.7}$$

As x^*/π is irrational, the Fourier coefficients a_n are determining uniquely and the uniqueness problem is solved.

Now we obtain the conditional stability estimate.

Theorem 2.1.2. Let $u(x,t) \in u(M)$ and solve the problem (1.0.1), (2.1.1), (2.1.2). If

$$|f(t)| \leq \varepsilon \tag{2.1.8}$$

and x^*/π be such that

$$\left| \frac{x^*}{\pi} - \frac{p}{q} \right| \geq \frac{\lambda(x^*/\pi)}{q^m},$$

where p, q are integers and m is natural, then

$$|u(x,t)| \leq C\varepsilon^{\frac{2}{m+2}\omega_0} M^{\frac{2}{m+2}(1-\omega_0)}, \tag{2.1.9}$$

where $0 < \omega_0 < 1$.

Proof. As $u(x,t) \subset u(M)$,then

$$|a_n| \leq 2Mn^{-3}. \tag{2.1.10}$$

From (2.1.8) it follows that

$$|F(p)| \leq \int_0^\infty e^{-\mathrm{Re}pt} \varepsilon dt = \frac{\epsilon}{\mathrm{Re}p}, \quad \mathrm{Re}p \geq p_0 > 0. \tag{2.1.11}$$

Denote by γ_n the following circle

$$\gamma_n = \left\{ p : |p + \frac{n^2}{n^2+1}| = \frac{1}{4}\left[\frac{(n+1)^2}{(n+1)^2+1} - \frac{n^2}{n^2+1}\right]\right\}.$$

In this circle, for $F(p)$ we have the following estimate

$$|F(p) \leq MCn. \tag{2.1.12}$$

In order then to obtain the estimate for $F(p)$ in the left half-plane by means of ε we shall make the following constructions. Consider the domain Q_n which is contained between the circle S_n with its center in the origin and the radius

$$R_n = \frac{3}{4}\frac{(n+1)^2}{(n+1)^2+1} + \frac{1}{4}\frac{n^2}{n^2+1}$$

and the circle B_n with the radius'

$$r_n = \frac{1}{2}\frac{(n+1)^2}{(n+1)^2+1} + \frac{1}{2}\frac{n^2}{n^2+1}.$$

The center of the circle B_n is in the real axis, so, that S_n and B_n are in contact in the point $\xi = (R_n, 0)$.

Construct an arc which intersects the domain Q_n in the first quadrant and the conformal mapping $(p - R_n)^{-1}$ maps this arc into the interval parallel to the real axis.

Denote by D_n the domain which is obtained from Q_n by removing its part from the first quadrant between the intersecting arc and the contacting point. In this domain D_n the function $F(p)$ in the intersecting arc E_n has the estimate (2.1.11), and in the remainder of the boundary it has limit estimate (2.1.12).

By the two-constant theorem in the domain D_n we have the estimate.

$$|F(p) \leq (\varepsilon p_0^{-1})^{\omega(p_n,E_n,D_n)}(MCn)^{1-\omega(p_n,E_n,D_n)}, \tag{2.1.13}$$

where $\omega(p_n, E_n, D_n)$ is the harmonic measure of E_n relative to the domain D_n in the point p_n.

For the one-connected domain D_n the conformal mapping $W = i(p - R_n)^{-1}$ transforms into the following semi-strip

$$R_n^{-1} < |JmW| < (2r_n - R_n)^{-1}, \text{ Re}W < b, \tag{2.1.14}$$

where constant $b > 0$. Consider the following conformal mapping

$$Z_n = \varphi(p_n, n)$$

$$= \frac{\pi}{2}\frac{R_n(2r_n - R_n)}{2(R_n - r_n)}\left\{W - i\left[\frac{1}{R_n} + \frac{1}{2}\left(\frac{1}{2r_n - R_n} - \frac{1}{R_n}\right)\right]\right\}$$

$$= i\frac{\pi}{4}\frac{R_n(2r_n - R_n) + r_n(p_n - R_n)}{(R_n - r_n)(R_n - p_n)}. \tag{2.1.15}$$

Transformation (2.1.15) maps semi-strip (2.1.14) into the semi-strip p

$$|JmZ| \leq \frac{\pi}{2}, \text{ Re}Z \leq b'. \tag{2.1.16}$$

Substituting the values R_n and r_n into (2.1.15) we may see that

$$Z_n = \varphi(p_n, n)$$

$$= \frac{a_0 n^{10} + a_1 n^9 + \ldots + a_{10} + p_n(a_0' n^{10} + a_1' n^9 + \ldots + a_{10}')}{b_0 n^{10} + b_1 n^9 + \ldots + b_{10} + p_n(b_0' n^{10} + b_1' n^9 + \ldots + b_{10}')}$$

$$= \frac{(a_0 - a_0')n^{10} + \ldots}{(b_0 - b_0')n^{10} + \ldots} = \frac{a_0 - a_0'}{b - b_0'} + O(\frac{1}{n}) = \tilde{Z}_n + O(\frac{1}{n}), \tag{2.1.17}$$

where $\tilde{Z}_n$ are constants and $Jm\tilde{Z}_n = 0$, $\mathrm{Re}Z < 0$.

It is well known that if the function $W(z)$ conformally maps a domain D onto D' and a set E maps onto E' then for the harmonic measure the following equality

$$\omega(Z, E, D) = \omega(W(z), E', D')$$

holds. Using this property, from (2.1.13) we obtain

$$|F(p)| \leq (\varepsilon p_0^{-1})^{\omega(Z_n, H, p)}(MCn)^{1-\omega(Z_n, H, p)}, \tag{2.1.18}$$

where $\omega(Z_n, H, p)$ is the harmonic measure of the end-wall H of the semi-strip (2.1.16) in the point Z_n. As it is known (see Evgrafov, 1968)

$$\omega(Z_n, H, p) = \frac{2}{\pi} Jm \ln \frac{1 + ie^{Z_n - b'}}{1 - ie^{Z_n - b'}}. \tag{2.1.19}$$

We must show that $\omega(Z_n, H, p)$ is estimated from below by positive number not depending on n. Note, that Z_n are situated in $\mathrm{Re}Z < 0$. The expression from (2.1.19) we transform as follows

$$\frac{1 - e^{2(Z_n - b')} + 2ie^{Z_n - b'}}{1 + e^{2(Z_n - b')}} = \frac{1 - e^{2(Z_n - b')}}{1 + e^{2(Z_n - b')}} + i\frac{2e^{Z_n - b'}}{1 + e^{2(Z_n - b')}}.$$

For the argument Ψ of the latter complex number we have

$$tg\Psi = \frac{2e^{Z_n - b'}}{1 - e^{2(Z_n - b')}}, \quad \Psi = \tan^{-1}\frac{2e^{Z_n - b'}}{1 - e^{2(Z_n - b')}}.$$

In this case

$$\omega(Z_n, H, p) = \frac{2}{\pi}\tan^{-1}\frac{2e^{Z_n - b'}}{1 - e^{2(Z_n - b')}} = \frac{2}{\pi}\tan^{-1}\frac{2}{e^{-(Z_n - b')} - e^{Z_n - b'}}$$

$$= \tan^{-1}\frac{2}{e^{b' - Z_n} - e^{-(b' - Z_n)}} = \frac{2}{\pi}\tan^{-1}\tan^{-1}\sinh^{-1}(b' - Z_n).$$

From (2.3.17) it follows that there exists such constant d, that the inequality

$$\sinh^{-1}(b' - Z_n) = \frac{1}{2}(e^{b'-Z_n} - e^{-b'-Z_n}) < e^{b'-Z_n} < e^{b'+d}.$$

holds. From here

$$\sinh^{-1}(b' - Z_n) > e^{-(b'+d)}.$$

Finally, by the monotonicity of $\tan^{-1}$ we have

$$\omega(Z_n, H, p) > \frac{2}{\pi}\tan^{-1} e^{-(b+d)} = \omega_0.$$

Now estimate (2.1.13) may be rewritten as follows

$$|F(p)| \leq (\varepsilon p_0^{-1})^{\omega(Z_n,H,p)}(MCn)^{1-\omega(Z_n,H,p)}$$

$$\leq n(\varepsilon p_0^{-1})^{\omega_0}(MCn)^{1-\omega_0} = n\varepsilon_0. \tag{2.1.20}$$

Formulae (2.1.7) and (2.1.20) yield

$$|a_n \sin nx^*| \leq \frac{|\Gamma_n| n\varepsilon_0}{2\pi}.$$

Note, that

$$|\Gamma_n| = 2\pi\frac{1}{2}\frac{(n^2+1)(n^2+2n+1) - n^2 - n^2(n^2+2n+1)}{(n^2+1)(n^2+2n+2)}$$

$$= \frac{(2n+1)\pi}{(n^2+1)(n^2+2n+2)} \leq \pi\varepsilon n^{-3}.$$

Consider the partial sum

$$\sum_{n=1}^{N} |a_n \sin nx^*| \leq \frac{1}{2\pi}\sum_{n=1}^{N} |\Gamma_n| n\varepsilon_0 \leq \frac{1}{2}\sum_{n=1}^{N} n^{-3} n\varepsilon_0 \leq C_2\varepsilon_0.$$

Further,

$$\sum_{n=1}^{N} |a_n|^2 |\sin nx^*|^2 \le \left(\sum_{n=1}^{N} |a_n \sin nx^*|\right)^2 \le C_2^2 \varepsilon_0^2. \qquad (2.1.21)$$

It follows from Theorem 2.1.2 condition that for each rational number pq^{-1} for the point x^*/π we have

$$\left|q\frac{x^*}{\pi} - p\right| \ge \frac{\lambda(x^*/\pi)}{q^{m-1}}. \qquad (2.1.22)$$

Due to the following relation

$$\left|\sin kx^*\right| = \left|\sin \frac{kx^*}{\pi}\pi\right| = \left|\sin\left\{\frac{kx^*}{\pi} - \pi - \left[\frac{kx^*}{\pi}\right]\pi + \left[\frac{kx^*}{\pi}\right]\pi\right\}\right|$$
$$|\sin\left\{\frac{kx^*}{\pi} - \left[\frac{kx^*}{\pi}\right]\right\}\pi|$$

and using inequality (2.1.22), we have

$$\left|\sin kx^*\right| = \left|\sin\left\{\frac{kx^*}{\pi} - \left[\frac{kx^*}{\pi}\right]\right\}\pi\right|$$
$$> \sin\frac{\lambda(\frac{x^*}{\pi})\pi}{k^{m-1}} \ge \frac{1}{N^m}, \quad k = 1,..,N.$$

Hence follows

$$|\sin kx^*| \ge \frac{1}{N^{2m}}. \qquad (2.1.23)$$

Inequalities (2.1.22), (2.1.23) yield

$$\sum_{n=1}^{N} |a_n|^2 \le \frac{N^{2m} C_2 \varepsilon_0^2}{4\pi^2 \lambda^2(x^*/\pi)}. \qquad (2.1.24)$$

Using *a priori* estimate (2.1.10) we obtain

$$\sum_{n=N+1}^{\infty} |a_n| \le 2M \sum_{n=N+1}^{\infty} \frac{1}{n^3} \le 2M \int_{N+1}^{\infty} \frac{dx}{x^3} \qquad (2.1.25)$$

Now, the Holder inequalities, (2.1.24), (2.1.25) yield

$$|u(x,t)| \leq \left|\sum_{k=1}^{\infty} a_k \sin kx\right| \leq \sum_{k=1}^{N} |a_k| + \sum_{k=N+1}^{\infty} |a_k|$$

$$\leq \left(\sum_{k=1}^{N} |a_k|^2\right)^{1/2} + \sum_{k=N+1}^{\infty} |a_k| \leq \frac{N^m C_2 \varepsilon_0}{2\pi\lambda(x^*/\pi)} + 2MN^{-2}. \quad (2.1.26)$$

We now find the value $\bar{N}$, which realizes the minimum of the right-hand side of (2.1.26)

$$\left[\frac{N^m C_2 \varepsilon_0}{2\pi\lambda(x^*/\pi)} + 3MN^{-2}\right] = \frac{mN^{m-1}C_2\varepsilon_0}{2\lambda(x^*/\pi)\pi} - \frac{6M}{N+3} = 0,$$

$$\bar{N} = \left[\frac{6M\pi\lambda(x^*/\pi)}{mC_2\varepsilon_0}\right]^{-\frac{2}{m+2}}.$$

Substituting this N into (2.1.26) and using (2.1.20) we obtain the following estimate

$$|u(x,t)| \leq c\varepsilon^{\frac{2}{m+2}\omega_0} M^{\frac{2}{m+2}(1-\omega_0)}.$$

Remark. Thus we may quarantee the Holder stability of the initial problem, if x^* belongs to the set $X(\sigma)$ with the following property

$$|\sin nx^2| \geq \lambda(x^*)n^{-\sigma}$$

for all $n = 1, 2, \ldots$,with some $\lambda(x^*) > 0$, $\sigma \geq 1$. Using certain results of the Diophantine approximation theory, Shishatskii (see Lavrent'ev *et al.*, 1980) had shown that $X(1) = C$, $\mu X(1) = 0$ and $X(1)$ is dense in $[0, \pi]$. If $\sigma > 1$, then $\mu X(\sigma) = \pi$. Here M is the Lebesque measure.

From the research of Shishatskii (see Lavrent'ev *et al.*, 1980 pp.72-73) it follows that problem (1.0.1), (2.1.1), (2.1.2) has the uniqueness property with the probability 1, and has the uniform stability estimate of the Holder type with the probability arbitrary close to 1.

§2.2 The inverse problem for the pseudo-parabolic equation without initial data.

In the domain

$$D = \{(x,t): \quad 0 \le x \le \pi, \quad 0 \le t \le T\}$$

consider the pseudoparabolic equation

$$(1 - L)u_t(x,t) = Lu(x,t), \tag{2.2.1}$$

where $L = \frac{\delta^2}{\delta x^2} - q(x)$.

Let be given the zero boundary conditions

$$u(0,t) = u(\pi,t) = 0 \tag{2.2.2}$$

be given, together with the following information about the solution

$$u_x(0,t) = f(t). \tag{2.2.3}$$

It is required to find an unknown coefficient $q(x)$ from equation (2.2.1) and the solution $u(x,t)$

Analogous problems were investigated earlier for parabolic equations for a certain initial condition. This problem, and analogous problems for the equations of parabolic type were formulated by Shishatskii.

The continuous solution to problem (2.2.1), (2.2.2) has the form

$$u(x,t) = \sum_{k=1}^{\infty} a_k e^{-\frac{\lambda_k t}{\lambda_k+1}} \varphi_k(x), \tag{2.2.4}$$

where λ_k are the eigenvalues, and $\varphi_k(x)$ are the normed eigenfunctions of L with boundary conditions (2.2.2).

Using condition (2.2.3) we have

$$f(t) = \sum_{k=1}^{\infty} a_k e^{-\frac{\lambda_k t}{\lambda_k+1}} \varphi'(0). \tag{2.2.5}$$

Theorem 2.2.1. Let all the eigenvalues λ_k of the operator L be positive and the conditions

$$1)\ 0 < \lambda_1 < \lambda_2 < \ldots < \lambda_n \quad 2)\ |a_k| \le C_k^{-3}, \quad \varphi_k(0) \ne 0, \forall k,$$

hold, where C_k is a constant. Then the inverse problem (2.2.1)- (2.2.3) has a unique solution : $u(x,t)$ and $q(x)$ in the class of the functions are even relative to the point $x = \pi/2$.

Proof. Since $\varphi_k'(0) \sim \sqrt{2/\pi k}$, the following estimate

$$|a_k \varphi_k'(0)| \le C_k^{-2}$$

holds. From here it follows that series (2.2.5) converges for $t \ge 0$ and its sum is bounded. Moreover, since the terms of series (2.2.5) are entire analytic functions of t and the series uniformly converges in the half-plane $\mathrm{Re}t \ge \sigma > 0$, then the function $f(t)$ has an analytic continuation into the half-plane $\mathrm{Re}t > 0$.

Now apply the Laplace transformation to (2.2.5)

$$\int_0^\infty \exp(-pt) \sum_{k=1}^\infty \frac{1}{\alpha_k} \exp(-\mu_k t) dt = F(p), \tag{2.2.6}$$

where

$$\alpha_k = \frac{1}{a_k \varphi_k'(0)}, \quad \mu_k = \lambda_k(\lambda_k + 1)^{-1}, \quad F(p) = \int_0^\infty \exp(-pt) f(t) dt.$$

Then, using (2.2.6) we have

$$\sum_{k=1}^\infty [\alpha_k(\mu_k + p)]^{-1} = F(p).$$

Knowing μ_k it is easy to find $\lambda_k = \mu_k(1-\mu_k)^{-1}$. Then, by the Levinson theorem (see Levinson, 1949), by the eigenvalues λ_k the function $q(x)$ is uniquely determined. By $q(x)$ we may determine the eigenfunctions $\varphi_k(x)$ corresponding to the eigenvalues λ_k of the operator L. Thus, we obtain the coefficient a_k and, therefore, the solution.

§2.3 The interior periodic problem

Consider again equation (1.0.1) in the domain

$$D = \{(x, t) : \quad 0 \le x \le \pi, \quad t \ge 0\}$$

with the zero boundary conditions (2.1.1).

Denote by L the polygon which lies in the semi-strip D with the successive vertices

$$(0,0), \quad (\pi, \pi/\alpha), \quad (0, 2\pi\alpha^{-1}), (\pi, 3\pi\alpha^{-1}), \ldots,$$

where α is a constant.

Suppose that we know the following information on solution of (1.0.1)

$$u(x,t)|_{(x,t)\in L} = g(t). \tag{2.3.1}$$

It is required to find by this information the conditions of unique determination of the solution to (1.0.1), and to obtain the stability estimate.

Evidently, the solution $u(x,t)$ may be continued in the whole axis $-\infty < x < \infty$ as a 2π-periodic function of x. The infomation on $u(x,t)$ in L is equivalent to the information on $u(x,t)$ in the line $x = \alpha t$. Practically, the information (2.3.1) may result in the transmitter moving by L. Or, the transmitter is making a reflexive-translational horizontal movement, and the probing domain is moving.

The continuous solution to problem (1.0.1), (2.1.1) from the class $u(M)$ (see Section 2.1) may be represented in the form

$$u(x,t) = \sum_{k=1}^{\infty} a_k e^{-\frac{k^2 t}{k^2+1}} \sin kx. \tag{2.3.2}$$

Denote L by $\alpha(t)$; then

$$\alpha(t - \frac{2\pi k}{\alpha}), \quad for \quad t \in \left[\frac{\pi}{\alpha} 2k, \ \frac{\pi}{\alpha}(2k+1)\right] \in B_1, \tag{2.3.3}$$

$$\pi - \alpha\left[t - \frac{\pi}{\alpha}(2k+1), \quad for \quad t \in [\frac{\pi}{\alpha}(2k+1)], \ \frac{\pi}{\alpha}(2k+2)\right] \in B_2, \tag{2.3.4}$$

Substituting (2.3.3), (2.3.4) in turn into (2.3.2) and taking into account (2.3.1), we obtain

$$g(t) = \begin{cases} u[\alpha(t - \frac{2\pi k}{\alpha}), t] = \sum\limits_{k=1}^{\infty} a_k e^{-\frac{k^2 t}{k^2+1}} \sin k\alpha t & \text{if } t \in B_1, \\ u[\pi - \alpha(t + \frac{2k+k}{\alpha}), t] = -\sum\limits_{k=1}^{\infty} a_k e^{-\frac{k^2 t}{k^2+1}} \sin k\alpha t & \text{if } t \in B_2, \end{cases}.$$

Determine $f(t)$ as follows

$$f(t) = \begin{cases} g(t) & \text{for } t \in B_1, \\ -g(t) & \text{for } t \in B_2. \end{cases}$$

If $g(t)$ is given, then $f(t)$ is also given, where for the function $f(t)$ the equality

$$f = \sum_{k=1}^{\infty} a_k e^{-\frac{k^2 t}{k^2+1}} \sin k\alpha t \tag{2.3.5}$$

holds for all t.

Theorem 2.3.1. The solution to problem (1.0.1), (2.1.1), (2.3.1) is unique in the class $u(M)$.

Proof. Represent (2.3.5) in the following form

$$f = \sum_{k=1}^{\infty} a_k e^{-\frac{k^2 t}{k^2+1}} \frac{e^{ikt\alpha} - e^{-ikt\alpha}}{2i}. \tag{2.3.6}$$

Apply to both sides of (2.3.6) the Laplace transformation

$$F(p) = \int_0^\infty f(t)e^{-pt}dt = \frac{1}{2i}\sum_{k=1}^{\infty}\int_0^\infty \Big\{\exp\left[-(p + k^2(k^2+1)^{-1} - ik\alpha)t\right]$$

$$- \exp[-(p + k^2(k^2+1)^{-1} + ik\alpha)t]\Big\}\, dt$$

$$\frac{1}{2i}\sum_{k=1}^{\infty} a_k \left(\frac{1}{p + k^2(k^2+1)^{-1} - ik\alpha} - \frac{1}{p + k^2(k^2+1)^{-1} + ik\alpha}\right). \tag{2.3.7}$$

Now integrate (2.3.7) by the circles Γ_n with centers in the points $p_n = -n^2(n^2+1)^{-1} + in_\alpha$ and the radius $1/4\beta$,where $\beta = \min(\alpha, 1)$. Then, by the integral Cauchy formula we obtain

$$\int_{\Gamma_n} F(p)dp = \pi a_n. \tag{2.3.8}$$

From here we obtain the Fourier coefficients a_n and, thus, the solution is unique.

Now obtain the estimate of solution.

Theorem 2.3.2. Let $u \in u(M)$ solves (1.0.1), (2.1.1), (2.3.1). If

$$g(t) \leq \varepsilon,$$

then

$$|u(x,t)| \leq C_1 \varepsilon^{C_2(\alpha)}, \tag{2.3.9}$$

where C_1 is a constant, $C_2(\alpha) > 0$.

Proof. From $g(t) < \varepsilon$ we have $f(t) < \varepsilon$. Therefore, the estimate

$$|F(p)| = \left|\int_0^\infty e^{-pt} f(t)dt\right| \leq \frac{\varepsilon}{\mathrm{Re}p} \tag{2.3.10}$$

holds, where $\mathrm{Re}p \geq p_0 > 0$.

From $u(x,t) \in u(M)$ follows

$$|a_n| \leq 2Mn^{-3}. \tag{2.3.11}$$

Estimate the function $F(p)$ outside the disk γ_n with the center $p_n = -n^2(n^2+1)^{-1} + in\alpha$ and the radius $\beta/8$

$$|F(p)| = \left| \frac{1}{2i} \sum_{k=}^{\infty} a_k \left(\frac{1}{p - p_k} - \frac{1}{p - \hat{p}_k} \right) \right|$$

$$= \left| \frac{1}{2i} \sum_{k=1}^{\infty} a_k \frac{p_k - \bar{p}_k}{(p - p_k)(p - \bar{p}_k)} \right| \leq \sum_{k=1}^{\infty} |a_k| \frac{C_k \alpha}{\beta} \leq \frac{C}{\beta},$$

where $\bar{p}_n = -n^2(n^2+1)^{-1} - in\alpha$.

Here we have used estimate (2.3.11) (nonessential constants we again denote by C).

Consider the circle γ with the center $p_0 = (1, \beta/2)$ and the radius $\beta/2$. Outside γ for $F(p)$ we have

$$|F(p)| \leq \frac{C}{\beta}. \tag{2.3.12}$$

To obtain the desired estimate we shall use the estimate (2.3.8). Therefore, we need to obtain the estimate of $F(p)$ by ε in Γ_n. Estimate $F(p)$ in the circle Γ_0 with the center p_0 and the radius $\beta/2$. This circle is most farthest the domain Re$p > 0$ of all Γ_n. That is why the estimate will be the worse of all. To solve this problem we shall make the following constructions.

Denote by D the domain which is obtained by the union of the disk Γ with the center p_0 and the radius $\beta/2$, and the rectangle p

$$0 \leq Jmp \leq \beta, \qquad |\text{Re}p| \leq 1.$$

Let γ_n' be the circle with the center p_0 and the radius $4\beta/q$. Choose in γ_n' the point $\tilde{p}$: Re$\tilde{p} = -1$, $\quad Jm\tilde{p} = 17\beta/q$. Now construct the angle $\lambda\pi$ which is based in the end-wall of D, the vertex lies in Γ and the bisectrix is parallel to Re$p = 0$ and contains the point $\tilde{p}$. In the sides of this angle estimate (2.3.12) holds, and in the end-wall, estimate (2.3.10). In this case, in the point $\tilde{p}$ by the Carleman lemma (see Stoilov, 1962) we have

$$|F(\tilde{p})| \leq (C\beta^{-1})^{1-(2R^{-1})^{-\lambda}} \varepsilon^{(2R^{-1})^{-\lambda}} = \varepsilon_1, \tag{2.3.13}$$

where r is the distance from the vertex $\lambda\pi$ to $\tilde{p}$, R being the maximal distance from the angle vertex to the end-wall. In our case

$$r = \sqrt{\frac{1}{4}\beta^2 - \frac{16}{81}\beta^2} = \frac{\sqrt{17}}{18}\beta,$$

$$R = \sqrt{(2+r)^2 + \left(\frac{\beta^2}{2} + \frac{4}{9}\right)^2} = \sqrt{4 + \frac{\sqrt{17}}{9}\beta + \frac{17}{18}\beta^2},$$

$$\lambda = \frac{2}{\pi}\tan^{-1}\frac{\beta}{36 + \sqrt{17}\beta}.$$

Estimate (2.3.13) holds for the points of the right semi-circle γ' , and in the left semi-circle γ' holds (2.3.12).

Finally, consider the domain Ω which is the disk γ' without γ_n. In each point of Ω and, in particular, in the point $\bar{p}$ of the circle $\Gamma_0(Re\bar{p} = -1\frac{1}{4}\beta, \quad Jm\bar{p} = \frac{1}{2}\beta)$, by the two constants theorem

$$F(\bar{p}) \leq \left(\frac{C}{\beta}\right)^{1-\omega} \varepsilon_1^{\omega},$$

where ω is the harmonic measure of the right semi-circle γ' relative to Ω calculated in the point $\bar{p}$. Since the transformation $z = \frac{1}{\beta}p$ is conformal and transfers Ω into the domain not depending on β, then ω does not depend on β (take into account that the harmonic measure is invariant relative to conformal mappings). So, $F(p)$ in Γ_0 and particularly in Γ has the estimate

$$|F(p)| \leq (C\beta^{-1})^{1-\omega}[(C\beta^{-1})^{1-(rR^{-1})^{-\lambda}} \varepsilon^{(rR^{-1})^{-\lambda}}]^{\omega} = \varepsilon_2. \tag{2.3.14}$$

From (2.3.8) using estimate (2.3.14) we may easily obtain the following relation for coefficients

$$|a_n| \leq \frac{1}{\pi} \int\limits_{\Gamma_n} F(p)dp \leq \frac{1}{\pi} \varepsilon_2 |\Gamma_n| = C\beta\varepsilon_2,$$

where $|\Gamma_n|$ is the length of the circle Γ_n.

Now estimate the solutions

$$|u(x,t)| \leq \sum_{k=1}^{n} |a_k| + \sum_{k=N+1}^{\infty} |a_k| = \frac{N\beta\varepsilon_2}{2} + \frac{2}{N^2}.$$

Minimizing by N we have

$$|u(x,t)| \leq c\varepsilon^{\frac{2}{3}} \varepsilon_2^{\frac{2}{3}}.$$

Using (2.3.14), represent the latter estimate in the form

$$u(x,t) \leq C\beta^{\frac{2}{3}} (C\beta^{-1})^{\frac{2}{3}(1-\omega)} (C\beta^{-1})^{\frac{2}{3}\omega[1-(2R^{-1})^{-\lambda}]} \varepsilon^{\frac{2}{3}\omega(rR^{-1})^{-\lambda}}.$$

Introduce the notation

$$C_1(\alpha) = c^{\frac{5}{3} - \frac{2}{3}\omega(rR^{-1})^{-\lambda}} \beta^{\frac{2}{3}\omega(rR^{-1})^{-\lambda}},$$

$$C_2(\alpha) = \frac{2}{3}\omega(rR^{-1})^{-\lambda}.$$

Then

$$|u(x,t)| \leq C_1(\alpha)\varepsilon^{C_2(\alpha)}.$$

Thus the theorem is proved.

§2.4 Recovering the solution to the pseudoparabolic equation by its values in the curve.

Let in the semi-strip

$$D = \{(x,t): \quad 0 \le x \le \pi, \quad t \ge 0\}$$

be given the solution to equation (1.0.1) with zero boundary conditions (2.1.1). Consider a continuous curve $\alpha(t)$ lying in D and assume that the solution to (1.0.1), (2.1.1) is given in this curve

$$u(\alpha(t), t) = f(t). \tag{2.4.1}$$

The question arises, when this information is sufficient to determine the function $f(t)$ in the semi-strip.

If $\alpha(t) = \text{const}$ then, generally speaking, the considered problem is not unique (see Lavrent'ev *et al.*, 1980)

It occurs that this case of non-uniqueness is in some sense exclusive. First, we shall prove the auxiliary lemma.

Lemma 2.4.1. Let a real and continuous in D function $\alpha(t)$ be such that $\lim \alpha(t)$ does not exist. Then for each $k = 1, 2, \dots$ there exists a sequence $\{t_n^{(k)}\}$ that for $n \to 0$.

$$t_n^{(k)} \to \infty, \qquad \exp\left[-\left(\frac{1}{k^2+1} - \frac{1}{(k+1)^2+1}\right) t_n^{(k)}\right] \\ \times \left|\sin k\alpha(t_n^{(k)})\right|^{-1} \to 0. \tag{2.4.2}$$

Proof. First we show that for all $k = 1, 2, \dots$ there exists such a sequence $\{t_n^{(k)}\}$ that for $n \to \infty$

$$t_n^{(k)} \to \infty, \quad \min_{m \in M} \left|\frac{1}{\pi}\alpha(t_n^{k)}) - \frac{m}{k}\right| \\ \times \exp\left[\left(\frac{1}{k^2+1} - \frac{1}{(k+1)^2+1}\right) t_n^{(k)}\right] \to \infty \tag{2.4.3}$$

(M is a set of all integers).

Let

$$\underline{\lim}_{t\to\infty} \alpha(t) < \overline{\lim}_{t\to\infty} \alpha(t).$$

Then the function $\frac{1}{\pi}\alpha(t)$ attains for $t \to \infty$ each number from the interval

$$(\frac{1}{\pi}\underline{\lim}_{t\to\infty}\alpha(t), \quad \frac{1}{\pi}\overline{\lim}_{t\to\infty}\alpha(t))$$

an infinite number of times. Let x_* be an irrational number from this interval and $\{t_n^{(k)}\}$ be such that $t_n^{(k)} \to \infty \quad (n \to \infty)$ and

$$\alpha(t_n^{(k)}) = \pi x^*.$$

Then

$$\min\left|\frac{1}{\pi}\alpha(t_n^{(k)}) - \frac{m}{k}\right| = C(k) > 0$$

and, therefore, (2.4.3) holds.

If $t_n^{(k)} \to \infty$ for $n \to \infty$ and (2.4.3) holds, then

$$\left|\sin k\alpha(t_n^k)\right| = \left|\sin[k\alpha(t_n^k) - m_0\pi]\right| = \sin\left|k\alpha(t_n^{(k)} - m_0\pi\right|$$

$$\geq \frac{2}{\pi}\left|k\alpha(t_n^{(k)}) - m_0\pi\right| = 2k\left|\frac{1}{\pi}\alpha(t_n^{(k)}) - \frac{m_0}{n}\right| = 2k\min_{m\in M}\left|\frac{1}{\pi}\alpha(t_n^k) - \frac{m}{k}\right|,$$

where m_0 is the nearest integer to $\pi^{-1}k\alpha(t_n^{(k)})$. This and (2.4.3) yield

$$\left|\sin k\alpha(t_n^{(k)})\right|\exp\left[\left(\frac{1}{k^2+1} - \frac{1}{(k+1)^2+1}\right)t_n^{(k)}\right]$$

$$\geq 2k\min_{m\in M}\left|\frac{1}{\pi}\alpha(t_n^k) - \frac{m}{k}\right|$$

$$\times\exp\left[\left(\frac{1}{k^2+1} - \frac{1}{(k+1)^2+1}\right)t_n^{(k)}\right] \to \infty, \quad n \to \infty.$$

Thus the lemma is proved.

Definition. We shall say that $u(x,t)$ belongs to $U(M)$ if u, u_γ are continuous in D and are uniformly bounded in this domain

$$|u|, |I_\gamma| \leq M,$$

where $\gamma = t, x, xx, xt, xxt$.

Theorem 2.4.1. Let a continuous function $\alpha \neq \text{const}$ satisfy the lemma condition. Then, from

$$f(t) = u(\alpha(t), t) \equiv 0, \quad u \in U(M),$$

it follows that $u(x,t) \equiv 0$ in D.

Proof. Solution $u(xt) \in U(M)$ to equation (1.0.1) with condition (2.1.1) may be represented in the form

$$u(x,t) = \sum_{k=1}^{\infty} a_k e^{(-1+\frac{1}{k^2+1})t} \sin kx.$$

By the theorem condition

$$\sum_{k=1}^{\infty} a_k e^{(-1+\frac{1}{k^2+1})t} \sin k\alpha(t) \equiv 0.$$

We shall show that from the latter equality follows that $a_k = 0$ for k=1,2, ...

$$\begin{aligned} a_1 e^{(-1+\frac{1}{k^2+1})t} \sin k\alpha(t) &= -\sum_{k=2}^{\infty} a_k e^{(-1+\frac{1}{k^2+1})t} \sin k\alpha(t) \\ &= -e^{(-1+\frac{1}{2^2+1})t} \sum_{k=2}^{\infty} a_k e^{-(\frac{1}{2^2+1}-\frac{1}{k^2+1})t} \sin \alpha(t). \end{aligned}$$

$$|a_1| \leq \frac{1}{|\sin \alpha(t)| \exp\left(\frac{1}{1^2+1} - \frac{1}{2^2+1}\right)} \sum_{k=2}^{\infty} |a_k| \exp\left[-\left(\frac{1}{2^2+1} - \frac{1}{k^2+1}\right)t\right],$$

. . . .

$$|a_k| \leq \frac{1}{|\sin k\alpha(t)| \exp\left[\left(\frac{1}{k^2+1} - \frac{1}{(k+1)^2+1}\right)t\right]}$$
$$\times \sum_{i=k+1}^{\infty} |a_i| \exp\left[-\left(\frac{1}{(k+1)^2+1} - \frac{1}{i^2+1}\right)t\right].$$

Now let $\{t_n^{(k)}\}$ be the sequence, constructed in the lemma; then

$$|a_k| \leq \frac{1}{|\sin \alpha(t_n^{(k)})| \exp\left[\left(\frac{1}{k^2+1} - \frac{1}{(k+1)^2+1}\right)t_n^{(k)}\right]}$$
$$\times \sum_{i=k+1}^{\infty} |a_i| \exp\left[-\left(\frac{1}{(k+1)^2+1} - \frac{1}{i^2+1}\right)t_n^{(k)}\right].$$

Going in this inequality to the limit for $n \to \infty$ we obtain

$$a_k = 0.$$

§2.5 The probability interpretation of multipoint problems.

The contents of this section correspond to the writings of Shishatskii.

Let in the domain $0 \leq x \leq \pi, \quad 0 \leq t < \infty$ be concentrated the thermal field

$$u(x,t) = \sum_{k=1}^{\infty} u_k e^{-k^2 t} \sin kx, \tag{2.5.1}$$

and let the measurement of this field in the point $x = a$ give

$$\sum_{k=1}^{\infty} u_k e^{-k^2 t} \sin ka = f(t). \tag{2.5.2}$$

The practical interest is in the problem of recovering the field $u(x,t)$ (or the sequence $\{u_k\}^{\infty}$) by the function $f(t)$. It is easy to show that solution to this problem is, generally speaking, not unique. If, for a example, $a/\pi = p/q, \quad k = nq$, where p, q, n are natural, then $\sin ka = \sin nq\pi = 0$. It denotes that the left side of (2.5.1) has no coefficients u_{nq} which, consequently, can not be obtained by the function $f(t)$. If a/π is irrational, then all u_k are determined by $f(t)$ uniquely. But in this case, u_k (and, therefore, the whole field $u(x,t)$) are essentially unstable relative to the $f(t)$ variations. Problems on the field recovery very often involve such properties. That is why it is very important to have such a setting that reduces its less desirable properies.

The idea of the suggested setting is simple. The point where the field is measured is a random variable uniformly distributed in the interval $(a - \delta, \ a + \delta)$.

In this case

$$\sum_{k=1}^{\infty} u_k e^{-k^2 t} \sin k\alpha, \quad 0 \le t < \infty, \tag{2.5.3}$$

is a random process and $f(t)$ is its realisation. The solution to the recovering field problem (2.5.1) we shall call the function $u(x,t)$ with the following coefficients u_k

$$\sum_{k=1}^{\infty} u_k e^{-k^2 t} \sin k\alpha = f(t) - \varphi(t), \tag{2.5.4}$$

where $\varphi(t)$ is the following random process

$$M\varphi(t) = 0, \quad D\varphi(t) \le \varepsilon^2,$$

This definition assumes that $f(t)$ is determined to within the random error with the dispersion $\le \varepsilon^2$.

The determination of $\{u_k\}^\infty$ from (2.5.4) is, evidently, not unique. But it turns out that the field variations arise because of this non--uniqueness and depend on ε-Holder, which in the ill-posed problem theory is entirely admissible.

Let this problem have two solution $u'(x,t), \quad u''(x,t)$. The equations (2.5.4) for them are as follows

$$\sum_{k=1}^{\infty} u'_k e^{-k^2 t} \sin k\alpha = f(t) - \varphi'(t),$$

$$\sum_{k=1}^{\infty} u''_k e^{-k^2 t} \sin k\alpha = f(t) - \varphi''(t),$$

where, denoting $V_k = u'_k - u''$, we have

$$\sum_{k=1}^{\infty} V_k e^{-k^2 t} \sin k\alpha = -\varphi'(t) + \varphi''(t),$$

$$D \sum_{k=1}^{\infty} V_k e^{-k^2 t} \sin k\alpha \leq 2\varepsilon^2,$$

or

$$\frac{1}{2\delta} \int_{a-\delta}^{a+\delta} \left[\sum_{k=1}^{\infty} V_k e^{-k^2 t} \sin k\alpha - \frac{1}{2\delta} \int_{a-\delta}^{a+\delta} \sum_{k=1}^{\infty} V_k e^{-k^2 t} \sin k\beta d\beta \right]^2 d\alpha \leq 2\varepsilon^2,$$

whence

$$\int_{a-\delta}^{a+\delta} \left[\sum_{k=1}^{\infty} V_k \left(\sin k\alpha - \frac{\sin k\delta}{k\delta} \sin k\alpha \right) e^{-k^2 t} \right]^2 d\alpha \leq 4\delta\varepsilon^2. \qquad (2.5.5)$$

Introduce the notations

$$V_k \left(\sin k\alpha - \frac{\sin k\delta}{k\delta} \sin ka \right) = \omega_k(\alpha),$$

$$\sum_{k=1}^{\infty} \omega_k(\alpha) e^{-k^2 t} = Z_\alpha(t).$$

The function $Z_\alpha(t)$ for each $\alpha \in [a-\delta, a+\delta]$ is analytic by t in the half-plane $\mathrm{Re}t > 0$ and is periodic by Jmt with the period 2π.

We want, by using (2.5.5), to evaluate $|Z_\alpha(t)|$ for $Jmt = 0$, $\mathrm{Re}t = 0$. To this purpose we have to evaluate the derivative

$$\frac{\partial}{\partial_\alpha}[Z_\alpha^2(t)]$$

for these t. Let, first, t be a real. Then

$$\frac{\partial}{\partial_\alpha}[Z_\alpha^2(t)] = 2\sum_{k=1}^{\infty} V_k\left(\sin k\alpha - \frac{\sin k\delta}{k\delta}\sin ka\right)e^{-k^2 t}$$
$$\times \sum_{k=1}^{\infty} kV_k \cos k\alpha e^{-k^2 t}.$$

Since

$$\max_{[a-\delta,a+\delta]}\left|\sin k\alpha - \frac{\sin k\delta}{k\delta}\sin ka\right| \leq \begin{cases} \sin k\delta & \text{if } k\delta < \pi/2, \\ 1 & \text{if } k\delta \geq \pi/2, \end{cases}$$

then

$$\left|\sum_{k=1}^{\infty} V_k\left(\sin k\alpha - \frac{\sin k\delta}{k\delta}\sin k\alpha\right)e^{-k^2 t}\right| \leq \sum_{k<\frac{\pi}{2\delta}} |V_k| \sin k\delta$$

$$+ \sum_{k<\frac{\pi}{2\delta}} |V_k| \leq 2M\left(\sum_{k<\frac{\pi}{2\delta}} \frac{\sin k\delta}{k^3} + \sum_{k\geq\frac{\pi}{2\delta}} k^{-3}\right)$$

$$\leq 2M\left(\delta\sum_{k<\frac{\pi}{2\delta}} \frac{\sin k\delta}{k\delta}\frac{1}{k^2} + \int_{\frac{\pi}{2\delta}-1}^{\infty} x^{-3}\right)dx$$

$$\leq 2M\left[\delta\sum_{k=1}^{\infty}\frac{1}{k^2} + \frac{1}{2}\frac{1}{(\frac{\pi}{2\delta}-1)^2}\right] = 2M\left[\frac{\pi}{6}\delta + \frac{2\delta^2}{(\pi-2\delta)^2}\right] \leq 4,7\delta M$$

if $\sigma \leq \frac{\pi}{4}$. And since

$$\left|\sum_{k=1}^{\infty} kV_k \cos k\alpha e^{-k^2 t}\right| \leq \sum_{k=1}^{\infty} k|V_k| \leq M\sum_{k=1}^{\infty}\frac{1}{k^2} = M\pi^2/6,$$

then

$$\frac{\partial}{\partial\alpha}Z_\alpha^2 \leq 16\delta M^2. \tag{2.5.6}$$

Here we obtain for $Jmt = 2\pi$

$$Z_\alpha^2(t) \leq E^2 = \begin{cases} 4\varepsilon + 16\delta^2 M^2, & \text{if } \delta < \frac{\varepsilon}{2M}, \\ 16\delta\varepsilon M, & \text{if } \delta \geq \frac{\varepsilon}{2M}. \end{cases}$$

On the other hand, for an arbitrary t with $\mathrm{Re}t \geq 0$ we have

$$|Z_\alpha^2(t)| = \left|\sum_{k=1}^{\infty} V_k \left(\sin k\alpha - \frac{\sin k\delta}{k\delta}\sin ka\right) e^{-k^2 t}\right|^2$$
$$\leq \left(\sum_{k<\pi/2\delta} |V_k| \sin k\delta + \sum_{k\geq\pi/2\delta} |V_k|\right)^2$$
$$\leq 4M^2\left[\frac{\pi^2}{6}\delta + \frac{2\delta^2}{(\pi-2\delta)^2}\right] < 22\delta^2 M^2.$$

For the coefficient $\omega_k(\alpha)$ we have the easily verified formula

$$\omega_k(\alpha) = \frac{1}{2\pi i}\int_p^{p+2\pi i} Z(t)e^{k^2 t}dt \tag{2.5.7}$$

In each point t of the semi-strip $\mathrm{Re}t \geq 0, 0 \leq Jmt \leq 2\pi$ for $z(t)$, due to the two constant theorem, we have

$$|Z(t)| \leq E^{1-\omega(t)}(\sqrt{22}\delta M)^{\omega(t)},$$

where $\omega(t)$ is the harmonic measure in the point t of the end-wall of the semi-strip. A fairly straightforward calculation gives

$$\omega(t) = \frac{2}{\pi} Jm \ln\frac{1-e^{-t/2}}{1+e^{-t/2}}.$$

Maximal value in the interval $[p, p+2\pi i]$ the function $\omega(t)$ attains in the point $p+\pi i$ and it is

$$\omega_p = \frac{2}{\pi}Jm\ln\frac{1+ie^{-p/2}}{1-ie^{-p/2}} = \frac{2}{\pi}\tan^{-1}\frac{2e^{-p/2}}{1-e^{-p}}.$$

Hence follows

$$|\omega(\alpha)| \leq E^{1-\omega_p}(\sqrt{22}\delta M)^{\omega_p}e^{k^2 p}.$$

Since

$$\max_{[a-\delta.a+\delta]}\left|\sin k\alpha - \frac{\sin k\delta}{k\delta}\sin ka\right| \begin{cases} \geq \sin^2\frac{k\delta}{2} & \text{for } k\delta < \pi/2, \\ \geq 1/2 & \text{for } \pi/2 \leq k\delta < \pi, \\ = 1 & \text{for } \pi \leq k\delta, \end{cases}$$

then

$$|V_k| \leq E^{1-\omega_p}(\sqrt{22}\delta M)^{\omega_p} e^{k^2 p} \cdot \begin{cases} \sin^{-2}\frac{k\delta}{2} & \text{for } k\delta < \pi/2, \\ 2 & \text{for } \pi/2 \leq k\delta < \pi, \\ 1 & \text{for } \pi \leq k\delta. \end{cases}$$

Chapter 3

Inverse problems for operator pseudoparabolic equations.

Inverse problems for pseudoparabolic equations were apparently first, considered by Atamanov and Mamayusupov (1990) by the methods suggested by Romanov (1984).

In this chapter we consider the inverse problems for operator pseudoparabolic and operator integro-differential equations. The theorems of existence and uniqueness are proved, and stability estimates are obtained.

Analogous problems for integro-differential equations for parabolic and hyperbolic type were considered by Grasseli *et al.* (1992), and by Lorenzi and Sinestrari (1986).

The methods suggested by Lorenzi *et al.* fail in our case besause of the specific character of the equation.

Let us introduce some notations. Let X be a Banach space with a norm $||\cdot||_X$, $T \in R, T > 0$. Then

1) $C([0,T];X) = \{u : [0,T] \to X, \quad u$ is a continuous function$\}$

$$||u||_c = \sup_{t\in[0,T]} ||u(x)||_x;$$

2)$C^1([0,T];X) = \{u : [0,T] \to X, \quad u$ is a continuously differentiable function$\}$.

$$||u(t)||_{C^1([0,T];X)} = \sup_{t\in[0,T]} ||u(t)||_X + \sup_{t\in[0,T]} ||u'(t)||_X.$$

3) $C_m([0,T];X)$ is the space of m-dimensional vector-functions with the elements from $C([0,T];X)$,

$$||u(t)||_{c,m} = \sup_{t\in[0,T]} \{\max_i ||u_i(t)||_X\},$$

where $u = (u_1, ..., u_m)$.

A^t is the transpose to the matrix A.

If X_1, X_2 are Banach spaces, then by $L(X_1, X_2)$ we denote the space of linear bounded operators with the norm sup

$$L(X_1, X_1) = L(X_1).$$

4) $C_m[0,t]$ is the space of m-dimensional vector-functions with the elements from $C[0,T]$,

$$||u(t)||_{C_m[0,T]} = \sum_{i=1}^{m} \sup_{t\in[0,T]} |u_i(t)|$$

$$u(t) = (u_1(t), ..., u_n(t))$$

5) $||x||, ||A||$ are the norms for m-dimensional vector and for the corresponding $m \times m$ matrix A.

§3.1 The inverse problem for operator pseudoparabolic equations.

Setting of the problem. Consider the following inverse problem

$$u'(t) = (Au)_t + a(t)(Au) + \sum_{i=1}^{m} C_i(t)B_iu(t) + f(t), \ t \in [0,T]. \quad (3.1.1)$$

$$u(0) = u_0, \quad u_0 \in D(A), \quad (3.1.2)$$

$$\Phi_i(Au - u) = g_i(t), \quad i = 1, .., m; \; t \in [0, T], \tag{3.1.3}$$

$$\Phi_i B_j u(t) = \alpha_{ij}(t), \quad i, j = 1, .., m; \; t \in [0, T], \tag{3.1.4}$$

where

$a(t), \alpha_{ij}$ are known functions from $C[0, T]$,

g_i are known functions from $C^1[0, T]$,

$f(t)$ is a known function from $C([0, T]; X)$,

$C_i(t)$ are unknown functions from $C[0, T]$,

$u(t)$ is an unknown function from $C^1([0, T], D(A))$, $i, j = 1, ..., m$.

So, the inverse problem is to find the functions $u(t)$, $C_i(t)$.

Assume that the following conditions hold true

I. Let $A : D(A) \subset X \to X$ be a closed linear operator, $B_i : D(B_i) \subset X \to X$ be closed linear operators, $i = 1, 2, ..., m$. Let, besides, $D(A) \subset D(B_i)$, $i = 1, 2, ...m$.

II. The operator $(A - I)$ is reversible, i.e.,

$$(A - I)^{-1} \in L(X).$$

III. Let

$$\Phi_i \in L(X; R),$$

$$\Phi_i B_j u(t) = \alpha_{ij}(t) \in C[0, T], \; i, j = 1, ..., m.$$

$$\det M(t) \equiv \begin{vmatrix} \alpha_{11}(t) & \alpha_{12}(t) & \dots & \alpha_{1m}(t) \\ \alpha_{21}(t) & \alpha_{22}(t) & \dots & \alpha_{2m}(t) \\ \dots & \dots & \dots & \dots \\ \alpha_{m1}(t) & \alpha_{m2}(t) & \dots & \alpha_{mm}(t) \end{vmatrix} \neq 0$$

for $t \in [0, T]$, $M(t) = (\alpha_{ij}(t))$.

Existence and uniqueness of solution. From (3.1.1)-(3.1.2) it follows that

$$(Au)_t + a(t)(Au) = u'(t) - \sum_{i=1}^{m} C_i(t)B_i u(t) - f(t),$$

$$(Au)\big|_{t=0} = Au_0.$$

Here,

$$Au = e^{-\int_0^t a(s)ds} Au_0 + \int_0^t e^{-\int_s^t a(\tau)d\tau}$$

$$\times \left[u'(s) - \sum_{i=1}^{m} C_i(s)B_i u(s) - f(s)\right] ds. \tag{3.1.5}$$

$$\int_0^t e^{-\int_s^t a(\tau)d\tau} u'(s)ds = u(t) - e^{-\int_0^t a(\tau)d\tau} u_0$$

$$- \int_0^t e^{-\int_s^t a(\tau)d\tau} a(s)u(s)ds. \tag{3.1.6}$$

Taking into account (3.1.5), from (3.1.6) it follows that

$$(A - I)u(t) = -\int_0^t e^{-\int_s^t a(\tau)d\tau} \Big[a(s)u(s)$$

$$+ \sum_{i=1}^{m} C_i(s)B_i u(s)\Big] ds + f_1(t), \tag{3.1.7}$$

where I is the identity operator,

$$f_1(t) = e^{-\int_0^t a(s)ds}(Au_0 - u_0) - \int_0^t e^{-\int_s^t a(\tau)d\tau} f(s)ds. \tag{3.1.8}$$

From (3.1.7) we have

$$u(t) = -(A-I)^{-1} \int_0^t e^{-\int_s^t a(\tau)d\tau} \Big[a(s)u(s)$$

$$+ \sum_{j=1}^m C_j(s)B_j u(s) \Big] ds + (A-I)^{-1} f_1(t). \qquad (3.1.9)$$

Operating by B_i from the left on equation (3.1.9) and introducing the notation

$$v_i(t) = B_i u(t), \quad i = 1, ..., m, \qquad (3.1.10)$$

we obtain

$$v_i(t) = -B_i(A-I)^{-1} \int_0^t e^{-\int_s^t a(\tau)d\tau} \Big[a(s)u(s)$$

$$+ \sum_{j=1}^m C_j(s)v_j(s) \Big] ds + B_i(A-I)^{-1} f_1(t), \quad i = 1, ..., m. \qquad (3.1.11)$$

Operating by A from the left on equation (3.1.9) we have

$$Au(t) = -A(A-I)^{-1} \int_0^t e^{-\int_s^t a(\tau)d\tau} \Big[a(s)u(s)$$

$$+ \sum_{j=1}^m C_j(s)B_j u(s) \Big] ds + A(A-I)^{-1} f_1(t). \qquad (3.1.12)$$

From here

$$(Au(t))_t = -A(A-I)^{-1} a(t)u(t) - \sum_{j=1}^m C_j(t)A(A-I)^{-1} B_j u(t)$$

$$+A(A-I)^{-1} \int_0^t e^{-\int_s^t a(\tau)d\tau} a(t) \Big[a(s)u(s)$$

$$+ \sum_{j=1}^m C_j(s)B_j u(s) \Big] ds + A(A-I)^{-1} f_1'(t). \qquad (3.1.13)$$

Substituting (3.1.12) and (3.1.13) into (3.1.1) we have

$$u'(t) = -A(A-I)^{-1}a(t)u(t) - \sum_{j=1}^{m} C_j(t)(A-I)^{-1}B_j u(t)$$
$$+A(A-I)^{-1}\int_0^t e^{-\int_s^t a(\tau)d\tau} a(\tau)\Big[a(s)u(s)$$
$$+\sum_{j=1}^{m} C_j(s)B_j u(s)\Big]ds + A(A-I)^{-1}f_1'(t)$$
$$-A(A-I)^{-1}\int_0^t e^{-\int_s^t a(\tau)d\tau} a(t)\Big[a(s)u(s)$$
$$+\sum_{j=1}^{m} C_j(s)B_j u(s)\Big]ds + a(t)A(A-I)^{-1}f_1(t) + f(t).$$

From here, taking into account (3.1.9), we obtain

$$u'(t) = A(A-I)^{-1}(A-I)^{-1}\int_0^t e^{-\int_s^t a(\tau)d\tau} a(t)$$
$$\times\Big[a(s)u(s) + \sum_{j=1}^{m} C_j(t)B_j u(s)\Big]ds$$
$$+\sum_{j=1}^{m} C_j(t)B_j u(s)\Big]ds - \sum_{j=1}^{m} C_j(t)(A-I)^{-1}B_j u(t) + f_2(t), \quad (3.1.14)$$

where

$$f_2(t) = A(A-I)^{-1}[f_1'(t) + a(t)f_1(t)] + f(t).$$

Using (3.1.8) and $(A-I)(A-I)^{-1} = I$ we have

$$f_2(t) = -(A-I)^{-1}f(t). \quad (3.1.15)$$

Operating by $\Phi_i(A-I)$ from the left on equation (3.1.14), and taking into account formula (3.1.5) and

$$\Phi_i(A-I)u'(t)=g_i'(t),$$

we have

$$g_i'(t)=\Phi_i A(A-I)^{-1}\int\limits_0^t e^{-\int\limits_s^t a(\tau)d\tau} a(t)\cdot\Big[a(s)u(s)$$

$$+\sum_{j=1}^m C_j(s)B_ju(s)\Big]ds-\sum_{j=1}^m C_j(t)[\Phi_iB_ju(t)]-\Phi_if(t). \qquad (3.1.16)$$

Let $M^{-1}(t)$ be the inverse to the matrix

$$M(t)=(\alpha_{ij}(t)),\quad M^{-1}(t)=(\beta_{ij}(t)).$$

Then, (3.1.16) yield

$$C(t)=M^{-1}(t)\Phi A(A-I)^{-1}\int\limits_0^t e^{-\int\limits_s^t a(\tau)d\tau} a(t)$$

$$\times\left[a(s)u(s)+\sum_{j=1}^m C_j(s)v_j(s)\right]ds+G(t). \qquad (3.1.17)$$

where

$$C(t)=\begin{pmatrix}C_1(t)\\ C_2(t)\\ \vdots\\ C_m(t)\end{pmatrix};\ \Phi=\begin{pmatrix}\Phi_1\\ \Phi_2\\ \vdots\\ \Phi_m\end{pmatrix};$$

$$G(t)=\begin{pmatrix}G_1(t)\\ G_2(t)\\ \vdots\\ G_m(t)\end{pmatrix};\ g(t)=\begin{pmatrix}g_1(t)\\ g_2(t)\\ \vdots\\ g_m(t)\end{pmatrix}. \qquad (3.1.18)$$

$$G(t)=-M^{-1}(t)[g'(t)+\Phi f(t)].$$

System (3.1.17) may be represented in the following form

$$C_i(t) = \sum_{j=1}^{m} \beta_{ij}(t)\Phi_j A(A-I)^{-1} \int_0^t e^{-\int_s^t a(\tau)d\tau} a(t)$$

$$\times \left[a(s)u(s) + \sum_{k=1}^{m} C_k(s)v_k(s)\right] ds + G_i(t), \quad i = 1, ..., m, \quad (3.1.19)$$

where

$$M^{-1}(t) = (\beta_{ij}(t)), \quad M(t) = (\alpha_{ij}(t)), \quad \alpha_{ij}(t) = \Phi_i B_j u(t),$$

$$G_i(t) = \sum_{j=1}^{m} \beta_{ij}(t)[g_j'(t) + \Phi_j f(t)]. \quad (3.1.20)$$

'he system of equations (3.1.9), (3.1.11), and (3.1.19) is equivalent to 'rse problem (3.1.1)-(3.1.4). Rewrite this system

$$u(t) = -(A-I)^{-1} \int_0^t e^{-\int_s^t a(\tau)d\tau} \Big[a(s)u(s)$$

$$+ \sum_{j=1}^{m} C_j(s)v_j(s)\Big] ds + (A-I)^{-1} f_1(t),$$

$$v_i(t) = -B_i(A-I)^{-1} \int_0^t e^{-\int_s^t a(\tau)d\tau} \Big[a(s)u(s)$$

$$+ \sum_{j=1}^{m} C_j(s)v_j(s)\Big] ds + B_i(A-I)^{-1} f_1(t),$$

$$C_i(t) = \sum_{j=1}^{m} \beta_{ij}(t)\Phi_j A(A-I)^{-1} \int_0^t e^{-\int_s^t a(\tau)d\tau} a(t)$$

$$\times \left[a(s)u(s) + \sum_{k=1}^{m} C_k(s)v_k(s)\right] ds + G_i(t), \quad i = 1, ..., m. \quad (3.1.21)$$

Conditions

$$D(A) \subset D(B_i), \ i = 1, .., m$$

and $D(A) = D(A-I)$ yield

$$A(A-I)^{-1},\ (A-I)^{-1},\ B_i(A-I)^{-1} \in L(X), i=1,...,m.$$

Introduce the space

$$Z = Z_1 \times Z_2 \times Z_3, \tag{3.1.22}$$

where

$$Z_1 = C([0,T];X); \quad Z_2 = C_m([0,T];X); \quad Z_3 = C_m[0,T]. \tag{3.1.23}$$

Denote

$$||\cdot||_1 = ||\cdot||_{C([0,T];X)}, \quad ||\cdot||_2 = ||\cdot||_{C,m},$$

$$||\cdot||_3 = ||\cdot||_{C_m[0,T]}, ||\cdot||_Z = ||\cdot||_1 + ||\cdot||_2 + ||\cdot||_3. \tag{3.1.24}$$

We shall stipulate the existence of unique fixed points for the mapping

$$S : Z \to Z,$$

$$S : (u,v,c) \to \Big(S_1(u,v,c),\ S_2(u,v,c),\ S_3(u,v,c)\Big),$$

$$v = \begin{pmatrix} v_1 \\ \vdots \\ v_m \end{pmatrix}, \ c = \begin{pmatrix} C_1 \\ \vdots \\ C_m \end{pmatrix}, \ S_2 = \begin{pmatrix} s_2^1 \\ \vdots \\ s_2^m \end{pmatrix}; \ S_3 = \begin{pmatrix} s_3^1 \\ \vdots \\ s_3^m \end{pmatrix},$$

$$S_1(u,v,c) = -(A-I)^{-1} \int_0^t e^{-\int_s^t a(\tau)d\tau} \Big[a(s)u(s)$$

$$+ \sum_{j=1}^m C_j(s)v_j(s)\Big] ds + (A-I)^{-1} f_1(t), \tag{3.1.25}$$

$$S_2^i(u,v,c) = -B_i(A-I)^{-1} \int_0^t e^{-\int_s^t a(\tau)d\tau} \Big[a(s)u(s)$$

$$+ \sum_{j=1}^m C_j(s)v_j(s)\Big] ds + B_i(A-I)^{-1} f_1(t), \ i=1,...,m. \tag{3.1.26}$$

$$S_3^i(u,v,c) = \sum_{j=1}^{m} \beta_{ij}(t)\Phi_j A(A-I)^{-1} \int_0^t e^{-\int_s^t a(\tau)d\tau} a(t)$$

$$\times \left[a(s)u(s) + \sum_{k=1}^{m} C_k(s)v_k(s)\right] ds + G_i(t), \quad i=1,...,m. \qquad (3.1.27)$$

Formulas (3.1.22), (3.1.8), and (3.1.20) show that such R exists that

$$||(A-I)^{-1}f_1(t)||_1 + ||B(A-I)^{-1}f_1(t)||_2 + ||G(t)||_3 \leq R, \quad (3.1.28)$$

where

$$B = \begin{pmatrix} B_1 \\ \vdots \\ B_m \end{pmatrix}.$$

Consider the ball

$$Z(R) = \{(u,v,c) \in Z : ||u||_1 + ||v||_2 + ||c||_3 \leq 2R\}. \qquad (3.1.29)$$

We show that for sufficiently small T, the map S transforms the ball $Z(R)$ into itself and will be a contracting mapping.

Define the functions

$$\Psi_1(t) = \int_0^t e^{-\int_s^t a(\tau)d\tau} a(s)u(s)ds, \qquad (3.1.30)$$

$$\Psi_2(t) = \sum_{k=1}^{m} \int_0^t e^{-\int_s^t a(\tau)d\tau} C_k(s)v_k(s)ds. \qquad (3.1.31)$$

For each $T' \in [0,T]$,formulas (3.1.30), (3.1.31) yield

$$||\Psi_1(t)||_X \leq NT'||u(t)||_1, \qquad (3.1.32)$$

$$||\Psi_2(t)||_X \leq NT'||c||_3||V||_2, \quad t \in [0,T]. \qquad (3.1.33)$$

where N is a known constant.

Using (3.1.28), (3.1.32), and (3.1.33), we may obtain from (3.1.25) the following

$$||S_1(u,v,c)||_1 \leq ||(A-I)^{-1} f_1(t)||_1$$

$$+ NT'[||u(t)||_1 + ||c||_3 \, ||v||_2]. \tag{3.1.34}$$

Analogously, (3.1.28), (3.1.32), (3.1.33), and (3.1.26) yield

$$||S_2(u,v,c)||_2 \leq ||B(A-I)^{-1} f_1(t)||_2$$

$$+ NT'[||u(t)||_1 + ||c||_3 \, ||v||_2]. \tag{3.1.35}$$

And finally, (3.1.28), (3.1.32), (3.1.33), and (3.1.27) yield

$$||S_3(u,v,c)||_3 \leq ||G(t)||_3 + NT'[||u(t)||_1 + ||c||_3 \, ||v||_2], \tag{3.1.36}$$

where $N > 0$ is a known constant depending on

$$||(A-I)^{-1}||_{L(X)}, \quad ||A(A-I)^{-1}||_{L(X)},$$

$$||B_i(A-I)^{-1}||_{L(X)}, \quad ||\Phi_i||_{L(X,R)}, \quad i = 1, ..., m.$$

Furthermore, N depends on the known functions and the matrices $a(t), M(t)$.

Combining (3.1.34), (3.1.35), and (3.1.36) we have

$$||S_1(u,v,c)||_1 + ||S_2(u,v,c)||_2 + ||S_3(u,v,c)||_3$$

$$\leq R + 3NT'(R+1)R \leq 2R \tag{3.1.37}$$

for sufficiently small $T' \in [0,T]$.

Now let (u_1, v^1, c^1), $(u_2, v^2, c^2) \in Z(R)$

$$S(u_1, v^1, c^1) - S(u_2, v^2, c^2) = (Z_1, Z^2, Z^3). \qquad (3.1.38)$$

Then, for $t \in [0, T]$, we obtain

$$Z_1(t) = -(A - I)^{-1} \int_0^t e^{-\int_s^t a(\tau)d\tau} \Big\{ a(s)[u_1(s) - u_2(s)]$$

$$+ \sum_{i=1}^m [C_i^1(s) - C_i^2(s)] v_i^1(s) + \sum_{i=1}^m C_i^2(s)[v_i^1(s) - v_i^2(s)] \Big\} ds, \quad (3.1.39)$$

$$Z_i^2(t) = -B_i(A - I)^{-1} \int_0^t e^{-\int_s^t a(\tau)d\tau} \Big\{ a(s)[u_1(s) - u_2(s)]$$

$$+ \sum_{i=1}^m [C_i^1(s) - C_i^2(s)] v_i^1(s) + \sum_{i=1}^m C_i^2(s)[v_i^1(s) - v_i^2(s)] \Big\} ds, \quad (3.1.40)$$

$$Z_i^3(t) = \sum_{j=1}^m \beta_{ij}(t) \Phi_j A(A - I)^{-1} \int_0^t e^{-\int_s^t a(\tau)d\tau} a(t) \Big\{ a(s)[u_1(s) - u_2(s)]$$

$$+ \sum_{k=1}^m [C_k^1(s) - C_k^2(s)] v_i^1(s) + \sum_{k=1}^m C_k^2(s)[v_k^1(s) - v_k^2(s)] \Big\} ds. \quad (3.1.41)$$

For each $T' \in [0, T]$, (3.1.39), (3.1.40), and (3.1.41) yield

$$||Z_1(t)||_1 + ||Z^2||_2 + ||Z^3(t)||_3 \leq 3NT' \Big[||u_1 - u_2||_1$$

$$+ 2R_m ||v^1 - v^2||_2 + 2R_m ||C^1 - C^2||_3 \Big]. \qquad (3.1.42)$$

Formula (3.1.42) shows that for sufficiently small $T' \in [0, T]$, the operator S will be a contractive mapping of the ball $Z(R)$.

Thus, we have obtained the following results:

Theorem 3.1.1. Let conditions I-III hold, then such $T' \in [0,T]$ exists that inverse problem (3.1.1)-(3.1.4) has a unique solution

$$\{u(t), C_1(t), ..., C_m(t)\}$$

from the space

$$C'([0,T'];\ D(A)) \times C_m[0,T'].$$

Really, we have proved the uniqueness of the solution to system (3.1.21) from the spaces

$$Z = C([0,T'];X) \times C_m([0,T'];X) \times C_m[0,T'].$$

Further, formulas (3.1.9), (3.1.18), (3.1.13) yield

$$u(t) \in C'([0,T];\ D(A))$$

Stability estimate. Let $M(t) = (\alpha_{ij}(t))$ be a matrix-function from condition III.

Definition. A matrix-function

$$M(t) = (\alpha_{ij}(t)) \in K(\beta)$$

if and only if

$$\sup_{t\in[0,T]} ||M^{-1}(t)|| \le \beta$$

where $\beta > 0$ is a constant not depending on $M(t)$.

It is clear that if

$$||M(t)x|| \ge \frac{1}{\beta}||x||, \quad \forall x \in R^n, \quad \forall t \in [0,T],$$

then $M(t) \in K(\beta)$.

Let $\{u_k(t), c^{(k)}(t)\}$, $k = 1,2$ be two solutions to inverse problems (3.1.1) - (3.1.4) with the data

$$f_k(t),\ u_{ok},\ g_i^k(t),\ \alpha_{ij}^{(k)}(t)$$

correspondingly from the spaces

$$C'([0,T];\ D(A)) \times C_m[0,T].$$

Then

$$u_k'(t) = (Au_k)_t + a(t)(Au_k) + \sum_{i=1}^{m} C_i^{(k)}(t)B_iu_k(t) = f_k(t),$$

$$t \in [0,T], \quad u_k(0) = u_{Ok},\ u_{ok} \in D(A),$$

$$\Phi_i(Au_k - u_k) = g_i^k(t), \quad i = 1,...,m,\ t \in [0,T],$$

$$\Phi_i B_j u_k(t) = \alpha_{ij}^{(k)}(t), \quad i,j = 1,...,m,\ k = 1,2,\ t \in [0,T]. \qquad (3.1.43)$$

Further to the conditions given on p. 54 we shall assume that

IV. $M_s(t) = (\alpha_{ij}^{(s)}(t)) \in K(\beta), \quad s = 1,2.$

Let us introduce the following notations

$$u_3(t) = u_1(t) - u_2(t),$$

$$c^3(t) = c^1(t) - c^2(t),$$

$$u_{03} = u_{01} - u_{02},$$

$$f_3(t) = f_1(t) - f_2(t),$$

$$\alpha_{ij}^3(t) = \alpha_{ij}^1(t) - \alpha_{ij}^2(t),$$

$$g_{i3}^3(t) = g_{i1}^1(t) - g_{i2}^2(t),$$

$$\Phi = \begin{pmatrix} \Phi_1 \\ \Phi_2 \\ \vdots \\ \Phi_m \end{pmatrix}, \; c^k(t) = \begin{pmatrix} C_1^k(t) \\ C_2^k(t) \\ \vdots \\ C_m^k(t) \end{pmatrix},$$

$$g_k(t) = \begin{pmatrix} g_{1k}(t) \\ g_{2k}(t) \\ \vdots \\ g_{mk}(t) \end{pmatrix}; \; v^k(t) = \begin{pmatrix} v_1^k \\ v_2^k \\ \vdots \\ v_m^k \end{pmatrix}, \tag{3.1.44}$$

$$v_i^k = B_i u_k, \quad i = 1, 2, ..., m.$$

Then, from (3.1.42) it follows that

$$u_3'(t) = (Au_3)_t + a(t)(Au_3) + \sum_{i=1}^{m} C_i^3(t) B_i u_1(t)$$

$$+ \sum_{i=1}^{m} C_i^2(t) B_i u_3(t) = f_3(t), \quad u_3(0) = u_{03} \in D(A),$$

$$\Phi_i[(A-I)u_3(t)] = g_i^3(t), \; \Phi_i B_j u_3(t) = \alpha_{ij}^3(t), \; i,j = 1, ..., m, \;\; t \in [0,T]. \tag{3.1.45}$$

By a similar series of operations, we obtain from (3.1.45) the following equivalent system

$$u_3(t) = -(A-I)^{-1} \int_0^t e^{-\int_s^t a(\tau)d\tau} \Big[a(s)u_3(s)$$

$$+ \sum_{i=1}^{m} C_i^3(s) B_i u_1(s) + \sum_{i=1}^{m} C_i^2(s) v_i^3(s) \Big] ds + (A-I)^{-1} f_3^1(t), \tag{3.1.46}$$

$$v_i^3(t) = -B_i(A-I)^{-1} \int_0^t e^{-\int_s^t a(\tau)d\tau} \Big[a(s)u_3(s)$$

$$+ \sum_{j=1}^{m} C_j^3(s) B_j u_1(s) + \sum_{j=1}^{m} C_j^2(s) v_j^3(s) \Big] ds + B_i(A-I)^{-1} f_3^1(t), \tag{3.1.47}$$

$$C^{(3)}(t) = M_1^{-1}(t)\Phi A(A-I)^{-1}\int_0^t e^{-\int_s^t a(\tau)d\tau} a(t)\Big[a(s)u_3(s)$$

$$+\sum_{j=1}^m C_j^3(s)B_j u_1(s) + \sum_{j=1}^m C_j^2(s)v_j^3(s)\Big]ds$$

$$- M_1^{-1}(t)[M_2(t) - M_1(t)]C^2(t) + G_3(t), \qquad (3.1.48)$$

where $M_2^{-1}(t)$ is the inverse matrix to $M_1(t) \in K(\beta)$,

$$f_3^1(t) = e^{-\int_s^t a(s)ds}(Au_{03} - u_{03}) - \int_0^t e^{-\int_s^t a(\tau)d\tau} f_3(s)ds, \qquad (3.1.49)$$

$$G_3(t) = -M_1^{-1}(t)[(g_3(t))'_t + \Phi f_3(t)]. \qquad (3.1.50)$$

Introduce the following notations

$$\Psi = ||f_3(t)||_{C([0,T];X)} + ||Au_{03} - u_{03}||_X + ||g_3'(t)||_{C_m[0,T]}$$

$$+ \sum_{i,j=1}^m ||\alpha_{ij}^3(t)||_{C[0,T]}. \qquad (3.1.51)$$

$$w_3(t) = ||u_3(t)||_X + \sum_{i=1}^m ||v_i^3(t)||_X + ||c^{(3)}(t)||, t \in [0,T]. \qquad (3.1.52)$$

For the positive fixed R_0 and r we define

$$Q_1(R_0) = \{u \in C^1([0,T]; D(A)) :$$

$$||B_i u(t)||_{C([0,T],X)} \leq R_0, \quad i = 1, ..., m\}, \qquad (3.1.53)$$

$$Q_2(r) = \{C(t) \in C_m[0,T] : ||C(t)||_{C_m[0,T]} \leq r, \} \qquad (3.1.54)$$

Further, we shall assume that

$$u_1(t), u_2(t) \in Q_1(R_0), \quad C^1(t), C^2(t) \in Q_2(r).$$

Taking into account (3.1.49), (3.1.50), (3.1.51), (3.1.44) we have

$$||(A-I)^{-1} f_3^1(t)||_{C([0,T];X)} + ||B_i(A-I)^{-1} f_3^1(t)||_{C([0,T];X)}$$

$$+\, ||M_1^{-1}[M_2(t) - M_1(t)]C^2(t) - G_3(t)||_{C_m[0,T]} \le q_0(\beta, R_0, r, T)\Psi, \tag{3.1.55}$$

where $q_0(\beta, R_0, r, T) > 0$ depends also on $a(t)$ and the norms

$$||(A-I)^{-1}||_{L(X)}, \quad ||B_i(A-I)^{-1}||_{L(X)}, \quad ||\Phi_i||_{L(X,R)}, \; i = 1, ..., m.$$

Taking into account (3.1.55) and (3.1.52), we obtain from (3.1.46), (3.1.47), and (3.1.48)

$$w(t) \le q_1 \int_0^t w(s)ds + q_0\Psi, \tag{3.1.56}$$

where $q_1 > 0$ is a constant depending on β, R_0, r, T, and

$$||(A-I)^{-1}||, \quad ||A(A-I)^{-1}||.$$

Applying the Gronwall-Bellman inequality to (3.1.56), we obtain

$$w(t) \le q\Psi, \quad t \in [0,T], \tag{3.1.57}$$

where $q > 0$ is a known constant depending on $a(t)$,

$$||(A-I)^{-1}||_{L(X)}, \quad ||A(A-I)^{-1}||_{L(X)}, \quad \beta, r, R_0, T,$$

$$||B_i(A-I)^{-1}||_{L(X)}, \quad ||\Phi_i||_{L(X,R)}, \quad i = 1, ..., m.$$

Taking into account (3.1.51), (3.1.52), (3.1.44), we obtain from (3.1.56)

$$||u_1(t)-u_2(t)||_{C([0,T];X)} \leq q\Big\{||f_1(t)-f_2(t)||_{C([0,T];X)}$$

$$+||A(u_{01}-u_{02})-(u_{01}-u_{02})||_X+\sum_{i=1}^{m}||g'_{i1}(t)-g'_{i2}(t)||_{C[0,T]}$$

$$+\sum_{i,j=1}^{m}||\alpha^1_{ij}(t)-\alpha^2_{ij}(t)||_{C[0,T]}\Big\}. \tag{3.1.58}$$

$$||v^1(t)-v^2(t)||_{C_m([0,T];X)} \leq q\Big\{||f_1(t)-f_2(t)||_{C([0,T];X)}$$

$$+||A(u_{01}-u_{02})-(u_{01}-u_{02})||_X+\sum_{i=1}^{m}||g'_{i1}(t)-g'_{i2}(t)||_{C[0,T]}$$

$$+\sum_{i,j=1}^{m}||\alpha^1_{ij}(t)-\alpha^2_{ij}(t)||_{C[0,T]}\Big\}. \tag{3.1.59}$$

$$||C^1(t)-C^2(t)||_{C_m[0,T]} \leq q\Big\{||f_1(t)-f_2(t)||_{C([0,T];X)}$$

$$+||A(u_{01}-u_{02})-(u_{01}-u_{02})||_X+\sum_{i=1}^{m}||g'_{i1}(t)-g'_{i2}(t)||_{C[0,T]}$$

$$+\sum_{i,j=1}^{m}||\alpha^1_{ij}(t)-\alpha^2_{ij}(t)||_{C[0,T]}\Big\}. \tag{3.1.60}$$

From (3.1.46) we have

$$Au_3(t)=-A(A-I)^{-1}\int_0^t e^{-\int_s^t a(\tau)d\tau}\Big[a(s)u_3(s)$$

$$+\sum_{i=1}^{m}C_i^3(s)B_iu_1(s)+\sum_{i=1}^{m}C_i^2(s)v_i^3(s)\Big]ds+A(A-I)^{-1}f_3^1(t), \tag{3.1.61}$$

$$(Au_3(t))'_t = -A(A-I)^{-1}\Big[a(t)u_3(t) + \sum_{i=1}^{m} C_i^3(t)B_iu_1(t)$$

$$+\sum_{i=1}^{m} C_i^2(t)v_i^3(t) + \int_0^t (A-I)^{-1}e^{-\int_s^t a(\tau)d\tau} a(t)[a(s)u_3(s)$$

$$+\sum_{i=1}^{m} C_i^3(s)B_iu_1(s) + \sum_{i=1}^{m} C_i^2(s)v_i^3(s)+\Big]ds$$

$$+ A(A-I)^{-1}(f_3^1(t))'_t, \tag{3.1.62}$$

$$(Au_3(t))'_t = Au'_3(t). \tag{3.1.63}$$

Taking into account estimates (3.1.58), (3.1.59), (3.1.60), and formulas (3.1.61), (3.1.62), we obtain the following estimate

$$||u_1(t) - u_2(t)||_{C([0,T];X)} + ||Au_1(t) - Au_2(t)||_{C([0,T],X}$$

$$+||u'_1(t) - u'_2(t)||_{C([0,T];X)} + ||Au'_1(t) - Au'_2(t)||_{C([0,T],X}$$

$$+\sum_{i=1}^{m} ||C_i^1(t) - C_i^2(t)||_{C[0,T]} \leq q\Big[||f_1(t) - f_2(t)||_{C([0,T];X)}$$

$$+||A(u_{01} - u_{02}||_X + ||u_{01} - u_{02}||_X$$

$$+\sum_{i=1}^{m} ||g'_{i1}(t) - g'_{i2}(t)||_{C[0,T]} + \sum_{i,j=1}^{m} ||\alpha_{ij}^1(t) - \alpha_{ij}^2(t)||_{C[0,T]}\Big]. \tag{3.1.64}$$

where $q > 0$ is a known constant depending on $\beta, r, R_0, T, a(t)$,

$$||(A-I)^{-1}||_{L(X)}, \quad ||A(A-I)^{-1}||_{L(X)},$$

$$||B_i(A-I)^{-1}||_{L(X)}, \quad ||\Phi_i||_{L(X,R)}, \quad i = 1, ..., m.$$

Thus, we have proved the following

Theorem 3.1.2. Let the conditions I-IV (pages 54, 62) hold, and

$$(u_1(t), C^1(t)), \quad (u_2(t), C^2(t))$$

be two solutions to inverse problem (3.1.1)-(3.1.4) from the class

$$C^1([0,T], D(A)) \times C_m[0,T].$$

Then stability estimate (3.1.64) holds.

Remark. Let Ω be an open bounded domain from R^n with a boundary Γ of the class C^2, and

$$a_{kh}, a_k, a_0, b_i \in C(\hat{\Omega}), \quad k,h = 1,...,n; \quad i = 1,...,m; \; a_0 \leq 1$$

for $x \in \Omega$.

Assume that

$$\sum_{k,h=1}^{n} a_{kh}(x)\xi_k\xi_h > 0$$

for each $x = (x_1, ..., x_n) \in \bar{\Omega}, \xi = (\xi_1, ..., \xi_n) \in R^n \setminus \{0\}$.

Let

$$Au(x) = \sum_{k,h=1}^{n} a_{kh}(x)u_{x_k x_h}(x) + \sum_{k=1}^{n} a_k(x)u_{x_k}(x) + a_0(x)u(x), \quad x \in \bar{\Omega},$$

$$B_i u(x) = b_i(x)(Au(x)), \quad x \in \bar{\Omega}, \; i = 1, ..., m_0 \leq m_1,$$

$$B_j u(x) = b_i(x)u(x), \quad x \in \bar{\Omega}, \; m_0 + 1 \leq j \leq m,$$

$$\Phi_i u(x) = u(x^i), \quad x^i \in \bar{\Omega}, \; i = 1, ..., m.$$

Consider the following inverse problem

$$u_t(t,x) = (Au(t,x))_t + a(t)(Au(t,x))$$

$$+\sum_{i=1}^{m} C_i(t)B_i u_1(t,x) = f(t,x), \quad (t,x) \in [0,T] \times \bar{\Omega}, \qquad (3.1.65)$$

$$u(0,x) = u_0(x), \quad x \in \bar{\Omega}, \qquad (3.1.66)$$

$$u(t,x) = 0, \quad (t,x) \in [0,T] \times \Gamma, \qquad (3.1.67)$$

$$\Phi[(A-I)u(t,x)] = \sum_{k,h=1}^{n} a_{kh}(x^i)u_{x_k x_h}(t,x^i) + \sum_{k=1}^{n} a_k(x^i)u_{x_k}(t,x^i)$$

$$+ [a_0(x^i) - 1]u(t,x^i) + g_i(t), \quad x^i \in \Omega, \quad i = 1,...,m. \qquad (3.1.68)$$

Here

$$\Phi[B_j u(t,x)] = \alpha_{ij}(t), \quad t \in [0,T], \ i = 1,...,m, \ j = 1,...,m. \quad (3.1.69)$$

It is clear that

$$\alpha_{ij}(t) = b_j(x^i)[g_i(t) + u(t,x^i)], \quad t \in [0,T],$$

for every $i = 1,...,m$; $j = 1,...,m_0$. And

$$\alpha_{ij}(t) = b_j(x^i)u(t,x^i), \quad t \in [0,T],$$

for every $i = 1,...,m$; $\quad j = m_0 + 1,...,m$.
Here $f(t,x)$ is a known function from

$$C([0,T];C(\bar{\Omega})), \quad g_i(t), \ i = 1,...,m.$$

The functions $g_i(t)$, $i = 1,...,m$ are certain functions from $C^1[0,T]$; α_{ij}, $a(t)$; $i,j = 1,..,m$ are the certain functions from $C[0,T]$; $u_0(x)$ is a known function; $u(t,x)$, $C_i(t)$, $i = 1,...,m$ are unknown functions.

In order to apply Theorems 3.1.1. and 3.1.2 to inverse problem (3.1.64) - (3.1.68) we define the corresponding spaces as follows

$$X = C(\bar{\Omega}),\ D(A) = \{u \in W^{2,p}(\Omega) : Au \in C(\bar{\Omega}),\ u = 0 \in \Gamma\},\ p > n.$$

It is clear that

$$D(B_i) = D(A),\ i = 1, ..., m_0,\ D(B_i) = X$$

for $i = m_0 + 1, ..., m$; $\bar{D}(A) \neq X$.

Assume that $u_0(X) \in D(A)$. By Lemmas 9,17 from Gilbarg and Trudinger (1989) and the imbedding theorem, the operator $(A - I)$ is reversible and

$$(A - I)^{-1} \in L(C(\bar{\Omega})).$$

Assume, besides, that

$$\det M(t) \equiv \begin{vmatrix} \alpha_{11}(t) & \alpha_{12}(t) & \ldots & \alpha_{1m}(t) \\ \alpha_{21}(t) & \alpha_{22}(t) & \ldots & \alpha_{2m}(t) \\ \ldots & \ldots & \ldots & \ldots \\ \alpha_{m1}(t) & \alpha_{m2}(t) & \ldots & \alpha_{mm}(t) \end{vmatrix} \neq 0$$

for each $t \in [0, T]$, $M(t) = (\alpha_{ij}(t))$.

Then the conditions I-III hold for inverse problem (3.1.65)-(3.1.69).

If there exists such $\beta \in (0, \infty)$ that for each $M(t) = (\alpha_{ij}(t)) \in R(\beta)$, i.e., for the inverse $M^{-1}(t)$ the estimate

$$\sup_{t \in [0,T]} ||M^{-1}(t)|| \leq \beta$$

holds, then the condition IV holds for inverse problem (3.1.65)-(3.1.69). In this case, the results of Theorems 3.1.1. and 3.1.2 hold true for inverse problem (3.1.65)-(3.1.69).

§3.2 The inverse problem for the operator integro- differential pseudoparabolic equation

Statement of the problem. We consider the following inverse problem

$$u_t'(t) = (Au)' + a(Au) + \sum_{m}^{i=1} \int_0^t K_i(t-s)B_i u(s)ds + f(t), \quad (3.2.1)$$

$$u(0) = u_0, \qquad u_0 \in D(A), \quad (3.2.2)$$

$$\Phi_i(A-I)u = g_i(t), \quad t \in [0,T],\ i = 1,\dots,m, \quad (3.2.3)$$

where $K_i(t)$ are unknown functions from $C[0,T]$, $i = 1,\dots,m$; $u(t)$ is an unknown function from $C^1([0,T]), D(A))$. Functions $f(t)$ from $C^1([0,T];X)$ and $g_i(t), i = 1,\dots,m$ from $C^2[0,T]$ are known.

Assume that the following conditions hold

I. Let $A : D(A) \subset X \to X$ be a closed linear operator; $B_i : D(B_i) \subset X \to X$ be closed linear operators and

$$D(A) \subset D(B_i), \qquad i = 1,\dots,m.$$

II. The operator $(A - I)$ has the inverse

$$(A-I)^{-1} \in L(X),$$

where I is a unit operator, $\Phi \in L(X,R), i = 1,\dots,m$.

III. $\det M(t) \neq 0$ for $t \in [0,T]$, where $M(t)$ is the following matrix

$$\begin{pmatrix} \Phi_1(B_i u_0) & \Phi_1(B_2 u_0) & \dots & \Phi_1(B_m u_0) \\ \Phi_2(B_1 u_0) & \Phi_2(B_2 u_0) & \dots & \Phi_2(B_m u_0) \\ \dots\dots & \dots\dots & \dots\dots & \dots\dots \\ \Phi_m(B_i u_0) & \Phi_m(B_2 u_0) & \dots & \Phi_m(B_m u_0) \end{pmatrix}$$

Assume that for a solution to inverse problem (3.2.1) - (3.2.3) the estimate

$$\|u(t)\|_C + \|u'(t)\|_C + \|\bar{K}(t)\|_{C_m[0,T]} \leq R_0 \tag{3.2.4}$$

holds, where $\bar{K}(t) = \{K_1(t), \dots, K_m(t)\}$; $R_0 > 0$ is known.

Represent equation (3.2.1) in the form

$$(Au)' + a(Au) = u'(t) - \sum_{i=1}^{m} \int_0^t K_i(t-s)B_i u(s)ds = f(s), \tag{3.2.5}$$

$$Au(0) = Au_0. \tag{3.2.6}$$

Differentiating (3.25) by t and introducing the notations

$$v(t) = u'(t) \tag{3.2.7}$$

we obtain

$$(Av)' + a(Av) = v'(t) - \sum_{i=1}^{m} \int_0^t K_i(t-s)B_i v(s)ds$$

$$- \sum_{i=1}^{m} K_i(t)B_i u_0 - f'(t), \tag{3.2.8}$$

$$v(0) = -(A-I)^{-1}[aAu_0 + f(0)], \tag{3.2.9}$$

$$Av(0) = -A(A-I)^{-1}[aAu_0 + f(0)]. \tag{3.2.10}$$

Determine Au from (3.2.5), (3.2.6)

$$Au = e^{-at}Au_0 + \int_0^t e^{a(t-s)}[u'(s)$$

$$-\sum_{i=1}^{m}\int_0^s K_i(s-\tau)B_iu(\tau)d\tau - f(s)]ds. \qquad (3.2.11)$$

We also have

$$\int_0^t e^{-a(t-s)}u'(s)ds = u(t) - e^{-at}u_0 - \int_0^t e^{-a(t-s)}au(s)ds. \qquad (3.2.12)$$

Taking into account (3.2.12) and reversing the operator $A - I$ from (3.2.10), we have

$$u(t) = -(A-I)^{-1}\int_0^t e^{-a(t-s)}[au(s)$$

$$+\sum_{i=1}^{m}\int_0^s K_i(s-\tau)B_iu(\tau)d\tau]ds + f_1(t), \qquad (3.2.13)$$

where

$$f_1(t) = e^{-at}u_0 - -(A-I)^{-1}\int_0^t e^{-a(t-s)}f(s)ds. \qquad (3.2.14)$$

Define Av from (3.2.8) and (3.2.10)

$$Av(t) = e^{-at}A(A-I)^{-1}[aAu_0 + f(0)] + \int_0^t e^{-a(t-s)}[v'(s)$$

$$-\sum_{i=1}^{m}\int_0^s K_i(s-\tau)B_iv(\tau)d\tau - \sum_{i=1}^{m}K_i(s)B_iu_0 - f'(s)]ds. \qquad (3.2.15)$$

Formula (3.2.12) and the reversion of the operator $A - I$ from (3.2.15) yield

$$v(t) = -(A-I)^{-1}\int_0^t e^{-a(t-s)}[av(s) + \sum_{i=1}^m K_i(s)B_i u_0$$

$$+\sum_{i=1}^m \int_0^s K_i(s-\tau)B_i v(\tau)d\tau]ds + f_2(t), \tag{3.2.16}$$

where

$$f_2(t) = -(A-I)^{-1}\Big\{e^{-at}[aAu_0 + f(0)] + \int_0^t e^{-a(t-s)} f'(s)ds\Big\}. \tag{3.2.17}$$

Introducing the notations

$$u_i(t) = B_i u(t), \qquad i = 1, \dots, m, \tag{3.2.18}$$

and operating by B_i on equation (3.2.13) from the left, we have

$$u_i(t) = -B_i(A-I)^{-1}\int_0^t e^{-a(t-s)}[au(s)$$

$$+\sum_m^{j=1} \int_0^s K_j(s-\tau)u_j(\tau)d\tau]ds + B_i f_1(t), \tag{3.2.19}$$

where $i = 1, \dots, m$.

Analogously, introducing the notations

$$v_i(t) = B_i v(t),\ i = 1, \dots, m \tag{3.2.20}$$

and operating by B_i on equation (3.2.16) from the left, we obtain

$$v_i(t) = -B_i(A-I)^{-1}\int_0^t e^{-a(t-s)}[av(s) + \sum_m^{j=1} K_j(s)B_j u_0$$

$$+\sum_m^{j=1} \int_0^s K_j(s-\tau)B_j v(\tau)d\tau]ds + B_i f_2(t), \tag{3.2.21}$$

where $i = 1, \dots, m$.

Taking into account notation (3.2.20), we obtain from (3.2.16)

$$Av(t) = -A(A-I)^{-1} \int_0^t e^{-a(t-s)}[av(s) + \sum_{i=1}^m K_i(s)B_j u_0$$

$$+ \sum_{i=1}^m \int_0^s K_i(s-\tau)v_i(\tau)d\tau]ds + Af_2(t). \qquad (3.2.22)$$

Differentiate this equation by t and use (3.2.22). As a result we have

$$(Av(t))' = -A(A-I)^{-1}[av(t) + \sum_{i=1}^m K_i(t)B_i u_0$$

$$+ \sum_{i=1}^m \int_0^t K_i(t-\tau)v_i(\tau)d\tau] + A(A-I)^{-1} \int_0^t e^{-a(t-s)} a[av(s)$$

$$+ \sum_{i=1}^m \int_0^s K_i(s-\tau)v_i(\tau)d\tau]ds + Af_2'(t). \qquad (3.2.23)$$

Substitute (3.2.22) and (3.2.23) into (3.2.8)

$$-A(A-I)^{-1}[av(t) + \sum_{i=1}^m K_i(t)B_i u_0 + \sum_{i=1}^m \int_0^t K_i(t-\tau)v_i(\tau)d\tau] + f_3(t)$$

$$= v'(t) - \sum_{i=1}^m \int_0^s K_i(t-\tau)v_i(\tau)d\tau - \sum_{i=1}^m K_i(t)B_i u_0, \qquad (3.2.24)$$

where

$$f_3(t) = Af_2'(t) + aAf_2(t) + f'(t) = -(A-I)^{-1}f'(t). \qquad (3.2.25)$$

Using the equality $(A-I)(A-I)^{-1} = I$ and (3.2.16) to (3.2.23), we have

$$(A-I)^{-1}[\sum_{i=1}^{m} K_i(t)B_iu_0+\sum_{i=1}^{m}\int_0^t K_i(t-s)v_i(s)ds]+v'(t)$$

$$-f_3(t)=aA-(A-I)^{-1}(A-I)^{-1}\int_0^t e^{-a(t-s)}[av(s)$$

$$+\sum_{i=1}^{m} K_iB_iu_0+\sum_{i=1}^{m}\int_0^s K_i(s-\tau)v_i(\tau)d\tau]ds-aA(A-I)^{-1}f_2(t). \quad (3.2.26)$$

Now, operate by $\Phi_i(A-I)$ from the left to equation (3.2.26) and use the formulae

$$\Phi_i(A-I)^{-1}v'(t)=g_i''(t),\quad i=1,\dots,m,\quad A(A-I)=(A-I)A. \quad (3.2.27)$$

As the result we obtain

$$\sum_{m}^{j=1}\Phi_i(B_ju_0)K_j(t)+\sum_{m}^{j=1}\Phi_i\int_0^t K_j(t-s)v_j(s)ds$$

$$=a\Phi_iA(A-I)^{-1}\int_0^t e^{-a(t-s)}[av(s)$$

$$+\sum_{m}^{j=1}K_j(s)B_ju_0+\sum_{m}^{j=1}\int_0^s K_j(s-\tau)v_j(\tau)d\tau]ds+\varphi_i(t), \quad (3.2.28)$$

where

$$\Phi_i(t)=\Phi_i[(A-I)f_3(t)-aAf_2(t)]-g_i''(t),\quad i=1,\dots,m. \quad (3.2.29)$$

Due to the condition III (p.), there exists the inverse matrix

$$M^{-1}(t)=\begin{pmatrix} \alpha_{11}(t) & \alpha_{12}(t) & \dots & \alpha_{1m}(t)\\ \alpha_{21}(t) & \alpha_{22}(t) & \dots & \alpha_{2m}(t)\\ \dots\dots & \dots\dots & \dots\dots & \dots\dots\\ \alpha_{m1}(t) & \alpha_{m2}(t) & \dots & \alpha_{mm}(t)\end{pmatrix} \quad (3.2.30)$$

to the matrix $M(t) = (\Phi_i(B_j u_0))_{i,j=1}^m$

Taking into account (3.2.30), from (3.2.28) we have

$$K_i(t) = \sum_m^{\nu=1} \alpha_{i\nu}(t)\Big\{ -\sum_m^{j=1} \Phi_\nu \int_0^t K_j(t-s)v_j(s)ds$$

$$= a\Phi_\nu A(A-I)^{-1} \int_0^t e^{-a(t-s)}[av(s)+$$

$$+\sum_m^{j=1} K_j(s)B_j u_0 + \sum_m^{j=1} \int_0^s K_j(s-\tau)v_j(\tau)d\tau]ds + \varphi_v(t)\Big\}, \quad (3.2.31)$$

where $i = 1, \dots, m$.

Systems (3.2.16), (3.2.21), (3.2.31) represent a closed system to determine the unknowns $v(t)$, $v_i(t)$, $K_i(t)$, $i = 1, \dots m$. Systems (3.2.13), (3.2.19) represent the closed system to determine the unknowns
$u(t), u_i(t)$, $i = 1, \dots, m$.

Thus, inverse problem (3.2.1) - (3.2.3) is equivalent to the following system

$$u(t) = -(A-I)^{-1}N_0(u, \bar{K}, \bar{u}) + f_1(t),$$

$$u_i(t) = B_i - (A-I)^{-1}N_0(u, \bar{K}, \bar{u}) + B_i f_1(t),$$

$$v(t) = -(A-I)^{-1}[N_0(v, \bar{K}, \bar{v}) + N_1(\bar{K})] + f_2(t), \quad (3.2.32)$$

$$v_j(t) = -B_j(A-I)^{-1}[N_0(v, \bar{K}, \bar{v}) + N_1(\bar{K})] + B_j f_2(t),$$

$$K_j(t) = \sum_{\nu=1}^{m} \alpha_{i\nu}(t)\Big\{ -\sum_{j=1}^{m} \Phi_v \int_0^t K_j(t-s)v_j(s)ds$$

$$+ a\Phi_\nu A(A-I)^{-1}[N_0(v, \bar{K}, \bar{v}) + N_1(\bar{K})] + f_v(t)\Big\} \quad (3.2.33)$$

where $i, j = 1, \ldots, m$, $\bar{K} = (K_1, \ldots, K_m)$, $\bar{u} = (u_1, ..., u)$, $\bar{v} = (v_1, ..., v_m)$,

$$N_0(u, \bar{K}, \bar{u}) = \int_0^t e^{-a(t-s)}[au(s) + \sum_{i=1}^m \int_0^s K_i(s-\tau)$$

$$\times u_i(\tau)d\tau]ds, N_1(\bar{K}) = \int_0^t e^{-a(t-s)}[\sum_{i=1}^m K_i(s)B_i u_0]ds. \quad (3.2.34)$$

The condition $D(A) \subset D(B_i)$, $i = 1, \ldots, m$ implies that $A(A-I)^{-1}$, $B_i(A-I)^{-1}$ are the elements from $L(X)$. Introduce the space

$$Z = Z_1 \cdot Z_2 \cdot Z_3 \cdot Z_4 \cdot Z_5,$$

where

$$Z_1 = Z_2 = C([0,T], X), \ Z_3 = Z_4 = C_m([0,T]; X),$$

$$Z_5 = C_m[0,T].$$

Introduce the notations

$$||\cdot||_1 = ||\cdot||_2 = ||\cdot||_C, \ ||\cdot||_3 = ||\cdot||_4 = ||\cdot||_{C,m},$$

$$||\cdot||_5 = ||\cdot||_{C_m[0,T]},$$

$||\cdot|| = \sum_{i=1}^5 ||\cdot||_i$ the norm in the space Z.

We shall show the existence of the unique fixed point of the map

$$S : Z \to Z,$$

$$S : z = \begin{pmatrix} u \\ v \\ \bar{u} \\ \bar{v} \\ \bar{K} \end{pmatrix} \to Sz = \begin{pmatrix} S_1 z \\ S_2 z \\ S_3 z \\ S_4 z \\ S_5 z \end{pmatrix},$$

$$\bar{u} = \begin{pmatrix} u_1 \\ \vdots \\ u_m \end{pmatrix}, \ \bar{v} = \begin{pmatrix} v_1 \\ \vdots \\ v_m \end{pmatrix}, \ S_3 z = \begin{pmatrix} S_3^1 z \\ \vdots \\ S_3^m z \end{pmatrix},$$

$$S_1 z = -(A - I)^{-1} N_0(u, \bar{K}, \bar{u}) + f_1(t), \tag{3.2.35}$$

$$S_2 z = -(A - I)^{-1} [N_0(v, \bar{K}, \bar{v}) + N_1(\bar{K})] + f_2(t), \tag{3.2.36}$$

$$S_3^i z = -B_i (A - I)^{-1} N_0(u, \bar{K}, \bar{u}) + B_i f_1(t), \tag{3.2.37}$$

$$S_4^i z = -B_i (A - I)^{-1} [N_0(v, \bar{K}, \bar{v}) + N_1(\bar{K})] + B_i f_2(t), \tag{3.2.38}$$

where $i, j = 1, \dots, m$, $\bar{K} = (K_1, \dots, K_m)$.

Taking into account (3.2.36), (3.2.37), (3.2.38), from (3.2.35) for sufficiently small $T' \in (0, T]$ we obtain

$$\|Sz\| \le R + C_2 T'(R + 1)R \le 2R, \tag{3.2.39}$$

where C_2 is a known positive constant not depending on

$$T' \in (0, T].$$

Now let

$$z_1 = (u^1, v^1, \bar{u}^1, \bar{v}^1, \bar{K}^1), \ z0_2 = (u^2, v^2, \bar{u}^2, \bar{v}^2, \bar{K}^2) \in Z(R)$$

Then for each $T' \in (0, T]$, from (3.2.32) and (3.2.33) we obtain

$$\|N_0(u^1, \bar{K}^1, \bar{u}^1) - N_0(\bar{u}^2, \bar{K}^2, \bar{u}^2)\|_X$$

$$\le C_0' T' [\|u' - u^2\|_1 + 2Rm\|\bar{u}^1 - \bar{u}^2\|_3 + 2Rm\|\bar{K}^1 - \bar{K}^2\|_5], \tag{3.2.40}$$

where C_0' is a known positive constant not depending on $T' \in (0, T]$. Applying the estimate

$$\|N_1(\bar{K}^1) - N_1(\bar{K}^2)\|_X \leq C_1 T' \|\bar{K}^1(t) - \bar{K}^2(t)\|_5 \qquad (3.2.41)$$

and estimate (3.2.40) to (3.2.35) we have

$$\|Sz_1 - Sz_2\| \leq C_3 T'(R+1)\|z_1 - z_2\|, \qquad (3.2.42)$$

where $C_3 > 0$ is a known constant not depending on $T' \in (0, T]$. and R.

From estimate (3.2.39) and (3.2.42) it follows that for sufficiently small $T' \in (0, T]$ the operator S maps the ball $Z(R)$ into itself and will be contractive.

Thus, we have solved the following results:

Theorem 3.2.1. Let the conditions I-IV (pp.) hold. Then there exists such $T' \in (0, T]$ that inverse problem (3.2.1)-(3.2.3) has a unique solution $\{u(t), K_1(t), ..., K_m(t)\}$ from the space $-C^1((0,T];\ D(A)) \times C_m[0,T]$ satisfying condition (3.2.4).

The stability estimate. Let $\{u^2(t), K_1^i(t), ..., K_m^i(t)\}$, $i = 1, 2$ be two solutions to inverse problem (3.2.1)-(3.2.3) with the data $f^i(t)$, u_{0k}, $g_{ij}(t)$ satisfying condition (3.2.4). Then

$$(u^i)' = (Au^i)' + a(Au^i) + \sum_{j=1}^{m} \int_0^t K_j^i(t-s) B_j u^i(s) ds + f^i(t),\ i = 1, 2, \qquad (3.2.43)$$

$$u^i(0) = u_{0i},\ u_{0i} \in D(A), \qquad (3.2.44)$$

$$\Phi(A - I)u^i = g_{ji}(t),\ t \in [0, T],\ j = 1, \ldots, m, \qquad (3.2.45)$$

$$\|u^i(t)\|_C + \|u^i(t)'\|_C + \|\bar{K}^i(t)\|_{C_m[0,T]} \leq R_0. \qquad (3.2.46)$$

Acting as above, from (3.2.43)-(3.2.46) we obtain the following system

$$z^i = Sz^i, \qquad i = 1, 2 \tag{3.2.47}$$

where

$$z^i = \begin{pmatrix} u^i \\ v^i \\ \bar{u}^i \\ \bar{v}^i \\ \bar{K} \end{pmatrix}, \ \bar{u}^i = \begin{pmatrix} u_1^i \\ \vdots \\ u_m^i \end{pmatrix}, \ \bar{v}^i = \begin{pmatrix} v_1^i \\ \vdots \\ v_m^i \end{pmatrix}, \ \bar{K}^i = \begin{pmatrix} K_1^i \\ \vdots \\ K_m^i \end{pmatrix},$$

$$S_2 z^i = \begin{pmatrix} S_1 z^i \\ S_2 z^i \\ S_3 z^i \\ S_4 z^i \\ S_5 z^i \end{pmatrix}, \ S_3 z^i = \begin{pmatrix} S_1^1 z^i \\ \vdots \\ S_3^m z^i \end{pmatrix},$$

$$S_1 z^i = -(A - I)^{-1} N_0(u^i, \bar{K}^i, \bar{u}^i) + f_1^i(t),$$

$$S_2 z^i = -(A - I)^{-1}[N_0(v^i, \bar{K}^i, \bar{v}^i) + N_1(\bar{K}^i)] + f_2^i(t),$$

$$S_3^n z^i = -B_n(A - I)^{-1} N_0(u^i, \bar{K}^i, \bar{u}^i) + B_n f_1^i(t),$$

$$S_4^n z^i = -B_n(A - I)^{-1}[N_0(v^i, \bar{K}^i, \bar{v}^i) + N_1(\bar{K}^i)] + B_n f_2^i(t)$$

$$S_5^n z^i = \sum_{\nu=1}^m \alpha_{n\nu}(t)\Big\{ -\sum_{j=1}^m \Phi_\nu \int_0^t K_j^i(t-s) v_j^i(s) ds$$

$$+ a\Phi_\nu A(A - I)^{-1}[N_0(v^i, \bar{K}^i, \bar{v}^i) + N_1(\bar{K}^i) + \varphi_\nu^i(t)\Big\}, \quad n = 1, \ldots, m. \tag{3.2.48}$$

The functions $f_1^i(t), f_2^i(t), f_3^i(t)$ are defined by formulas (3.2.14), (3.2.17), (3.2.15) correspondingly for $u_0 = u_{0i}$. The functions $f(t) = f^i(t), i = 1, 2,$ φ_ν^i are defined by formula (3.2.29) for

$$f_3(t) = f_3^i(t),\ f_2(t) = f_2^i(t),\ g_\nu(t) = g_{\nu i}(t), i = 1, 2.$$

From (3.2.33) and (3.2.34) we obtain

$$\|N_0(u^1, \bar{K}^1, \bar{u}^1) - N_0(u^2, \bar{K}^2, \bar{u}^2)\|_X \leq M_0 \int_0^t \Big\{ \|u^1(s)$$

$$- u^2(s)\|_X + \sum_{i=1}^m [|K_i^1(s) - K_i^2(s)| + \|u_i^1(s) - u_i^2(s)\|_X] \Big\} ds, \quad (3.2.49)$$

$$\|N_1(\bar{K}^1) - N_1(\bar{K}^2)\|_X \leq M_1 \int_0^t [\sum_{i=1}^m |K_i^1(s) - K_i^2(s)|] ds, \quad (3.2.50)$$

where M_0, M_1 are known positive constants depending on R_0, a, T.

Taking into account (3.2.48), (3.2.49), and (3.2.50), from (3.2.47) we have

$$w \leq M_2 \int_0^t w(s) ds + M_3,\ t \in [0, T], \quad (3.2.51)$$

where $M_2 > 0$ is a known constant depending on R_0, a, T, and on

$$\|(A - I)^{-1}\|_{L(X)},\ \|B_n(A - I)^{-1}\|_{L(X)},\ \|A(A - I)^{-1}\|_{L(X)},$$

$$\|\Phi_\nu\|_{L(X,R)},\ \sup_{t \in [0,T]} |\alpha_{n\nu}(t)|,\ n, \nu = 1, \ldots, m.$$

$$w(t) = \|u^1(t) - u^2(t)\|_X + \|v^1(t) - v^2(t)\|_X + \sum_{j=1}^m \Big\{ \|u_j^1(t) - u_j^2(t)\|_X$$

$$+ \|v_j^1(t) - v_j^2(t)\|_X + |K_j^1(t) - K_j^2(t)| \Big\},\ t \in [0, T], \quad (3.2.52)$$

$$M_3 = M_4\Big\{\|f_1^1(t) - f_1^2(t)\|_C + \|f_2^1(t) - f_2^2(t)\|_C$$

$$+\sum_{j=1}^{m}[\|B_j(f_1^1(t) - f_1^2(t))\|_C + B_j|f_2^1(t) - f_2^2(t))\|_C$$

$$+ \sup_{t\in[0,T]} |\varphi_j^1(t) - \varphi_j^2(t)|. \qquad (3.2.53)$$

M_4 is a known positive constant depending on

$$\sup_{t\in[0,T]} |\alpha_{n\nu}(t)|,\ n, \nu = 1, ..., m.$$

By formulas (3.2.14), (3.2.17) we have

$$\|f_1^1(t) - f_1^2(t)\|_C \leq M_0'[\|u_{01} - u_{02}\|_X + \|f^1(t) - f^2(t)\|_C] \quad (3.2.54)$$

$$\|B_j\Big(f_1^1(t) - f_1^2(t)\Big)\|_C \leq M_1'[\|B_j u_{01} - B_j u_{02}\|_X$$

$$+ \|f^1(t) - f^2(t)\|_C],\ j = 1, \ldots, m, \qquad (3.2.55)$$

$$\|f_2^1(t) - f_2^2(t)\|_C \leq M_2'[\|Au_{01} - Au_{02}\|_X$$

$$+ \|f^1(t) - f^2(t)\|_C] + \|(f^1(t))' - (f^2(t))'\|_C, \qquad (3.2.56)$$

$$B_j(\|f_2^1(t) - f_2^2(t))\|_C \leq M_3'[\|Au_{01} - Au_{02}\|_X$$

$$+\|f^1(t) - f^2(t)\|_C + \|(f^1(t))' - (f^2(t))'\|_C],\ j = 1, \ldots, m, \quad (3.2.57)$$

where M_0', M_1', M_2', M_3' are known positive constants.

Taking into account (3.2.17), (3.2.25), from (3.2.29) we obtain

$$\sup_{t\in[0,T]} |\varphi_j^1(t) - \varphi_j^2(t)| \leq M_4'[\|(f^1(t))' - (f^2(t))'\|_C$$

$$+ ||Au_{01} - Au_{02}||_x + ||f^1(t) - f^2(t)||_c + \|g_{j1}''(t) - g_{j2}''(t)\|_{C[0,T]}],\ (3.2.58)$$

$j = 1, \ldots, m$ and $M_4' > 0$ is a known constant depending on

$$\|A(A-I)^{-1}\|_{L(X)}, \ \|\Phi_j\|_{L(X,R)}, a_i.$$

Applying the Gronuoll-Bellman inequality, from (3.2.51) we obtain

$$w(t) \leq M_3 e^{M_2 T}, t \in [0,T]. \tag{3.2.59}$$

Combining (3.2.7), (3.2.18), (3.2.20), (3.2.53), and estimates (3.2.54)-(3.2.58), from (3.2.59) we obtain the stability estimate in the whole

$$\|u^1(t) - u^2(t)\|_{C^1} + \sum_{i=1}^{m} [\|B_i u^1(t) - B_i u^2(t)\|_{C^1}$$

$$+\|K_i^1(t) - K_i^2(t)\|_{C[0,T]}] \leq q\Big\{\|u_{01}$$

$$-u_{02}\|_X + \|Au_{01} - Au_{02}\|_X + \sum_{i=1}^{m}\Big[B_i u_{01} - B_i u_{02}\|_X$$

$$+ \|g_{i1}'' - g_{i2}''(t)\|_{C[0,T]} + \|f^1 - f^2\|_{C^1}\Big]\Big\}, \tag{3.2.60}$$

where $q > 0$ is a constant depending on R_0, a, T and on

$$\|(A-I)^{-1}\|_{L(X)}, \|A(A-I)^{-1}\|_{L(X)}, \|B_i(A-I)^{-1}\|_{L(X)},$$

$$\|\Phi_i\|_{L(X,R)}, \ i = 1, \ldots, m.$$

Thus, we have solved the following theorem.

Theorem 3.2.2. Let conditions I-IV hold, $(u^1(t), \bar{K}^1(t)$ and $(u^2(t), \bar{K}^2(t)$ be two solutions to inverse problem (3.2.1)-(3.2.3) from $C'([0,T]; D(A)) \times C_m[0,T]$ satisfying condition (3.2.46). Then the stability estimate (3.2.60) holds.

Remark. Let Ω be a bounded opened domain from R^n with the boundary Γ of the class C^2;

$$a_{kh}, a_k, a_0, b_1 \in C(\bar{\Omega}),\ k,h = 1,...,n,\ i = 1,...,m,\ a_0(x) \leq 1$$

for $x \in \bar{\Omega}$.

Assume, in addition, that

$$\sum_{k,h=1}^{n} a_{kh}(x)\xi_k\xi_h > 0$$

for each $x = (x_1, ..., x_n) \in \bar{\Omega}$ and $\xi = (\xi_1, ..., \xi_n) \in R^n \setminus \{0\}$.

Let

$$Au(x) = \sum_{k,h=1}^{n} a_{kh}(x)u_{x_k x_h}(x) + \sum_{k=1}^{n} a_k(x)u_{x_k}(x) + a_0(x)u(x),\ x \in \bar{\Omega},$$

$$B_i u(x) = b_i(x)(Au(x)), x \in \bar{\Omega},\ i = 1, \ldots, m_0 \leq m,$$

$$B_j u(x) = b_j(x)u(x), x \in \bar{\Omega},\ j = m_0 + 1, \ldots, m,$$

$$\Phi_i u(x) = u(x^i), x^i \in \Omega,\ i = 1, \ldots, m.$$

Consider the following inverse problem

$$u_t(t,x) = (Au(t,x))_t + a(Au(t,x)) \tag{3.2.61}$$

$$+\sum_{i=1}^{m} \int_0^t K_i(t-s)B_i u(s,x)ds + f(t,x), (t,x) \in [0,T] \cdot \bar{\Omega},$$

$$u(0,x) = u_0(x),\ x \in \bar{\Omega}, \tag{3.2.62}$$

$$u(t,x) = 0,\ (t,x) \in [0,T] \cdot \Gamma, \tag{3.2.63}$$

$$\Phi_i[(A-I)u(t,x)] \equiv \sum_{k,h=1}^{n} a_{kh}(x^i)u_{x_k x_h}(t,x^i)$$

$$+\sum_{k=1}^{n} a_k(x^i)u_{x_k}(t,x^i) + [a_0(x^i)-1]u(t,x^i) =$$

$$= g_i(t),\ t\in[0,T],\ i=1,\ldots,m, \tag{3.2.64}$$

where $f(x,t)$ is a given function from $C^1([0,T];C((\bar{\Omega}))$, $g_i(t)$ are given functions from $C^2[0,T]$; $u_0(x)$ is given. The functions $u(t,x)$ and $K_i(t)$, $i=1,...,m$ are unknowns.

In order to apply Theorems 3.2.1. and 3.2.2. to inverse problem (3.2.61)-(3.2.64) determine the corresponding space as follows

$$X = C(\bar{\Omega}), D(A) = \left\{u \in W^{2,p}(\Omega) : Au \in C(\bar{\Omega}), u = 0in\{\Gamma\},\ p>n.\right.$$

Assume that $u_0(x) \in D(A)$. By Lemma 9.17. from Gilbarg and Trudinger (1989) and the imbedding theorem the operator $(A-I)$ has the inverse $(A-I)^{-1} \in L(C(\bar{\Omega}))$. Assume, besides, that $\det M(t) \neq 0$ for $t\in[0,T]$, where $M(t)$ is the following matrix

$$\begin{pmatrix} \Phi_1(B_1u_0(x)) & \Phi_1(B_2u_0(x)) & \ldots & \Phi_1(B_mu_0(x)) \\ \Phi_2(B_1u_0(x)) & \Phi_2(B_2u_0(x)) & \ldots & \Phi_2(B_mu_0(x)) \\ \cdots\cdots & \cdots\cdots & \cdots\cdots & \cdots\cdots \\ \Phi_m(B_1u_0(x)) & \Phi_m(B_2u_0(x)) & \ldots & \Phi_m(B_mu_0(x)) \end{pmatrix}.$$

Then the conditions I-III hold for inverse problem (3.2.61)-(3.2.64). Therefore, for inverse problem (3.2.61)-(3.2.64) the results of Theorems 3.2.1. and 3.2.2. hold.

If $B_iu(x) = b_i(x)u(x)$, $i=1,...,m$, $x\in\bar{\Omega}$, then, $M(t)$ is the following matrix

$$\begin{pmatrix} b_1(x^1)u_0(x^1) & b_2(x^1)u_0(x^1) & \ldots & b_m(x^1)u_0(x^1) \\ b_1(x^2)u_0(x^2) & b_2(x^2)u_0(x^2) & \ldots & b_m(x^2)u_0(x^2) \\ \cdots\cdots & \cdots\cdots & \cdots\cdots & \cdots\cdots \\ b_1(x^m)u_0(x^m) & b_2(x^m)u_0(x^m) & \ldots & b_m(x^m)u_0(x^m) \end{pmatrix}.$$

§3.3 Recovering the kernels of the integro-differential operator of pseudoparabolic equation.

Statement of the problem. Consider the following inverse problem

$$u'_t(t) = (Au)'_t + a(Au) + \sum_{i=1}^{m_0} \int_0^t K_i(t-s)B_i u(s)ds$$

$$+ \sum_{i=1}^{m_1} \int_0^t C_i(s)E_i(s)ds + f(t),\ t \in [0,T], \tag{3.3.1}$$

$$u(0) = u_0,\ u_0 \in D(A), \tag{3.3.2}$$

$$\Phi_i u(t) = g_i(t),\ i = 1, ..., m = m_0 + m_1 \tag{3.3.3}$$

where $K_i(t), C_i(t)$ are unknown functions from $C[0,T]$, $i = 1, ..., m_0$, $j = 1, ..., m_1$. $u(t)$ is an unknown function, $f(t)$ is a known function from $C^1([0,T];X)$, $g_i(t)$ are known functions from $C^2[0,T]$, $i = 1, ..., m_0$; a is a given constant.

Assume that the following conditions hold

I. $A : D(A) \subset X \to X$ is a closed operator,

$B_i : D(B_i) \subset X \to X$ are closed linear operators,

$$D(A) \subset D(B_i), i = 1, ..., m_0,$$

$$E_j : D(E_j) \subset X \to X$$

are linear closed operators;

$$D(A) \subset D(E_j), j = 1, ..., m_1$$

II. The operator $(A - I)$ has the inverse and

$$(A - I)^{-1} \in L(X),$$

where I is a unit operator;

$$\Phi_i \in L(X, R), i = 1, ..., m_0 + m_1 = m.$$

III. Let for each

$$u(t) \in C^1([0,T]), D(A)),$$

$$\Phi_i u(t) = g_i(t),$$

$$\Phi_i (A - I)^{-1} E_j u(t) = \alpha_{ij} g_i(t),$$

$$t \in [0,T], i = 1, ..., m; \; j = 1, ..., m,$$

where α_{ij} are given constants.
IV. For $u_0 \in D(A)$, det $M(t) \neq 0$ for $t \in [0,T]$, where, $M(t)$ is the following matrix (we have denoted $A^* = (A - I)^{-1}$)

$$\begin{pmatrix} \Phi_1(A)^* B_1 u_0 & ... & \Phi_1(A)^* B_{m_0} u_0 \alpha_{11} g_1(t) & ... & \alpha_{1m} g_1(t) \\ \Phi_2(A)^* B_1 u_0 & ... & \Phi_2(A)^* B_{m_0} u_0 \alpha_{21} g_2(t) & ... & \alpha_{2m_1} g_2(t) \\ ... & ... & ... & ... & ... \\ \Phi_m(A)^* B_1 u_0 & ... & \Phi_m(A)^* B_{m_0} u_0 \alpha_{m1} g_m(t) & ... & \alpha_{mm_1} g_m(t) \end{pmatrix}.$$

Equation (3.3.1) may be represented as the following

$$(Au)'_t + a(Au) = u'_t - \sum_{i=1}^{m_0} \int_0^t K_i(t-s) B_i u(s) ds$$

$$- \sum_{i=1}^{m1} \int_0^t C_i(s) E_i u(s) ds - f(t). \tag{3.3.4}$$

$$Au(0) = Au_o \tag{3.3.5}$$

Introduce the notations

$$v(t) = u'_t(t), u(t) = \int\limits_0^t v(s)ds + u_0 \tag{3.3.6}$$

Differentiating (3.3.4) by t and taking into account (3.3.1), (3.3.2), and (3.3.6), we have

$$(Av_t)' + a(Av) = v'_t(t) - \sum_{i=1}^{m_o} \int\limits_0^t K_i(t-s)B_i v(s)ds$$

$$- \sum_{i=1}^{m_0} K_i(t)B_i u_0 - \sum_{i=1}^{m_1} C_i(t)E_i u(t) - f'_t(t), \tag{3.3.7}$$

$$v(0) = -A(A-I)^{-1}[aAu_0 + f(0)]. \tag{3.3.8}$$

$$Av(0) = -(A-I)^{-1}[aAu_0 + f(0)], \tag{3.3.9}$$

From (3.3.4) and (3.3.5) determine Au:

$$Au = e^{-at}Au_0 + \int\limits_0^t e^{-a(t-s)}[u'_s$$

$$-\sum_{i=1}^{m} \int\limits_0^s K_i(s-\tau)B_i u(\tau)d\tau$$

$$-\sum_{i=1}^{m_1} \int\limits_0^s C_i(\tau)E_i u(\tau)d\tau - f(s)]ds. \tag{3.3.10}$$

But

$$\int\limits_0^t e^{-a(t-s)}u'_s ds = u(t) - e^{-at}u_0$$

$$- \int\limits_0^t e^{-a(t-s)}au(s)ds. \tag{3.3.11}$$

Taking into account (3.3.11), from (3.3.10) we have

$$(A-I)u(t) = -\int_0^t e^{-a(t-s)}[au(s) + \sum_{i=1}^{m_0}\int_0^s K_i(s-\tau)B_iu(\tau)d\tau$$

$$+\sum_{i=1}^{m_1}\int_0^s C_i(\tau)E_iu(\tau)d\tau]ds + f_1(t), \qquad (3.3.12)$$

where I is a unit operator in X,

$$f_1(t) = e^{-at}(Au_0 - u_0) - \int_0^t e^{-a(t-s)}f(s)ds. \qquad (3.3.13)$$

From (3.3.12)we have

$$u(t) = -(A-I)^{-1}\int_0^t e^{-a(t-s)}\Big[au(s)$$

$$\sum_{i=1}^{m_0}\int_0^s K_i(s-\tau)B_iu(\tau)d\tau + \sum_{i=1}^{m_1}\int_0^s C_i(\tau)E_iu(\tau)d\tau\Big]ds$$

$$+(A-I)^{-1}f_1(t). \qquad (3.3.14)$$

From (3.3.7) and (3.3.9) define Av:

$$Av(t) = e^{-at}A(A-I)^{-1}[aAu_0 + f(0)$$

$$+\int_0^t e^{-a(t-s)}[v_s' - \sum_{i=1}^{m_0}\int_0^s K_i(s-\tau)B_iv(\tau)d\tau$$

$$-\sum_{i=1}^{m_0}B_{u_0} - \sum_{i=1}^{m_1}C_i(s)E_iu(s) - f_s'(s)]ds. \qquad (3.3.15)$$

Taking into account (3.3.11), from (3.3.15) we obtain

$$(A-I)v(t) = -\int_0^t e^{-a(t-s)}[av(s) + \sum_{i=1}^{m_0} K_i(s)B_i u_0$$

$$+\sum_{i=1}^{m_0}\int_0^s K_i(s-\tau)B_i v(\tau)d\tau + \sum_{i=1}^{m_1} C_i(s)E_i u(s)ds + f_2(t), \qquad (3.3.16)$$

where

$$f_2(e) = -e^{-at}[aAu_0 + f(0)] - \int_0^t e^{-a(t-s)} f_s'(s)ds. \qquad (3.3.17)$$

From (3.3.16) we have

$$v(t) = (A-I)^{-1}\int_0^t e^{-a(t-s)}[av(s) + \sum_{i=1}^{m_0} K_i(s)B_i u_0$$

$$+\sum_{i=1}^{m_0}\int_0^s K_i(s-\tau)B_i v(\tau)d\tau + \sum_{i=1}^{m_1} C_i(s)E_i u(s)ds$$

$$+ (A-I)^{-1} f_2(t). \qquad (3.3.18)$$

Using the operators B_i and E_j from the left on equation (3.2.14) and introducing the notations

$$u_i(t) = B_i u(t), \qquad i = 1, ..., m_0, \qquad (3.3.19)$$

$$w_j(t) = E_j u(t), \qquad j = 1, ..., m_1, \qquad (3.3.20)$$

we obtain

$$u_i(t) = -B(A-I)^{-1}\int_0^t e^{-a(t-s)}\Big[au(s)$$

$$+\sum_{k=1}^{m_0}\int_0^s K_k(s-\tau)u_k(\tau)d\tau + \sum_{k=1}^{m_1}\int_0^s C_k(\tau)w_k(\tau)d\tau\Big]ds$$

$$+ B_i(A-I)^{-1} f_1(t),\ i = 1, ..., m_0, \qquad (3.3.21)$$

$$w_j(t) = -E_j(A-I)^{-1}\int_0^t e^{-a(t-s)}\Big[au(s)$$

$$+\sum_{k=1}^{m_0}\int_0^s K_k(s-\tau)u_k(\tau)d\tau + \sum_{k=1}^{m_1}\int_0^s C_k(\tau)w_k(\tau)d\tau\Big]ds$$

$$+ E_j(A-I)^{-1}f_1(t),\ j=1,...,m_1. \tag{3.3.22}$$

Operating by B_i from the left on equation (3.3.18) and introducing the notations

$$v_i(t) = B_i v(t),\ i=1,...,m_0, \tag{3.3.23}$$

we obtain

$$v_i(t) = -B_j(A-I)^{-1}\int_0^t e^{-a(t-s)}\Big[av(s) + \sum_{k=1}^{m_0} K_k(s)B_k u_0$$

$$+\sum_{k=1}^{m_0}\int_0^s K_k(s-\tau)B_k v(\tau)d\tau + \sum_{k=1}^{m_1} C_k(s)E_k u(s)\Big]ds$$

$$B_i(A-I)^{-1}f_2(t),\ i=1,...,m_0. \tag{3.3.24}$$

Applying formulae (3.3.18), (3.3.20), and (3.3.23) to equation (3.3.7), we have

$$\sum_{j=1}^{m_0} K_j(t)B_j u_0 + \sum_{j=1}^{m_1}(E_j u(t))C_j(t)$$

$$= -\sum_{j=1}^{m_0}\int_0^t K_j(t-s)v_j(s)ds + a(A-I)^{-1}\int_0^t e^{-a(t-s)}\Big[av(s)$$

$$+\sum_{j=1}^{m_0} K_j(s)B_j u_0 + \sum_{j=1}^{m_0}\int_0^s K_j(s-\tau)v_j(\tau)d\tau$$

$$+\sum_{j=1}^{m_1} C_j(s)w_j(s)\Big]ds - f_t'(t) - a(A-I)^{-1}f_2(t)$$

$$- [(A-I)v(t)]'_t - a(A-I)v(t). \qquad (3.3.25)$$

Operating by $\Phi_i(A-I)^{-1}$, $i = 1,...,m$ from the left on equation (3.3.25), taking into account the condition III and notations (3.3.7), we obtain

$$\sum_{j=1}^{m_0} \Phi_i(A-I)^{-1}(B_j u_0)K_j(t) + \sum_{j=1}^{m_0} \alpha_{ij} g_i(t) C_j(t)$$

$$= -\sum_{j=1}^{m_0} \Phi_i(A-I)^{-1} \int_0^t K_j(t-s)v_j(s)ds + a\Phi_i(A-I)^{-1}(A-I)^{-1}$$

$$\times \int_0^t e^{-a(t-s)}[av(s) + \sum_{j=1}^{m_0} K_j(s)B_j u_0 + \sum_{j=1}^{m_0} \int_0^s K_j(s-\tau)v_j(\tau)d\tau$$

$$+ \sum_{j=1}^{m_1} C_j(s)w_j(s)]ds + f_{1i}(t), \qquad (3.3.26)$$

where $i = 1,...,m;\ m = m_0 + m_1$,

$$f_{1i}(t) = -\Phi_i(A-I)^{-1}[f'_t(t) + a(A-I)^{-1}f_2(t)] - g''_i(t) - ag'_i(t). \quad (3.3.27)$$

From the conditions III-IV there exists the inverse matrix

$$M^{-1}(t) = \begin{pmatrix} \beta_{11}(t) & \beta_{12}(t) & ... & \beta_{1m}(t) \\ \beta_{21}(t) & \beta_{22}(t) & ... & \beta_{2m}(t) \\ ... & ... & ... & \\ \beta_{m1}(t) & \beta_{m2}(t) & ... & \beta_{mm}(t) \end{pmatrix} \qquad (3.3.28)$$

to the matrix $M(t)$ (see condition IV).

Therefore,

$$K_i(t) = \sum_{\nu=1}^{m} \beta_{i\nu}(t)\Big\{ - \Phi_\nu \sum_{j=1}^{m_0}(A-I)^{-1}\int_0^t K_j(t-s)v_j(s)ds$$

$$+a\Phi_\nu(A-I)^{-1}(A-I)^{-1}\int_0^t e^{-a(t-s)}\Big[av(s)$$

$$+\sum_{j=1}^{m_0} K_j(s)B_j u_0 + \sum_{j=1}^{m_0}\int_0^s K_j(s-\tau)v_j(\tau)d\tau$$

$$+\sum_{j=1}^{m_1} C_j(s)w_j(s)\Big]ds + f_{1,\nu}(t)\Big\}, \qquad (3.3.29)$$

$$C_k(t) = \sum_{\nu=1}^{m} \beta_{k+m_0,\nu}(t)\Big\{ - \Phi_\nu \sum_{\nu=1}^{m_0}(A-I)^{-1}\int_0^t K(t-s)v_j(s)ds$$

$$+a\Phi_\nu(A-I)^{-1}(A-I)^{-1}\int_0^t e^{-a(t-s)}\Big[av(s)$$

$$+\sum_{j=1}^{m_0} K_j(s)B_j u_0 + \sum_{j=1}^{m_0}\int_0^s K_j(s-\tau)v_j(\tau)d\tau$$

$$+\sum_{j=1}^{m_1} C_j(s)w_j(s)\Big]ds + f_{1,\nu}(t)\Big\}, \quad i = 1,...,m_0;\ k = 1,...,m_1. \qquad (3.3.30)$$

Systems (3.3.14), (3.3.18), (3.3.21), (3.3.22), (3.3.24), (3.3.20) and (3.3.30) determine the closed system for finding the unknowns $u(t), v(t)$, $u_i(t)$, $w_j(t,)$ $v_i(t)$, $K_i(t)$, $C_j(t)$, $i = 1,...,m_0$, $j = 1,...,m_1$.

Thus, inverse problem (3.3.1) - (3.3.3) is equivalent to the following system

$$u(t) = -(A-I)^{-1}N_0(u,\bar{K},\bar{C},\bar{u},\bar{w}) + (A-I)^{-1}f_1(t),$$

$$u_j(t) = -B_i(A-I)^{-1}N_0(u,\bar{K},\bar{C},\bar{u},\bar{w}) + B_i(A-I)^{-1}f_1(t),$$

$$w_j(t) = -E_j(A-I)^{-1}N_0(u,\bar{K},\bar{C},\bar{u},\bar{w}) + E_j(A-I)^{-1}f_1(t),$$

$$v(t) = -(A-I)^{-1}N_1(v,\bar{K},\bar{C},\bar{v},\bar{w}) + (A-I)^{-1}f_1(t),$$

$$v_i(t) = -B(A-I)^{-1}N_1(v,\bar{K},\bar{C},\bar{v},\bar{w}) + B_i(A-I)^{-1}f_2(t),$$

$$K_i(t) = \sum_{m}^{\nu=1} \beta_{i\nu}(t)\Big\{ -\Phi_\nu(A-I)^{-1}\sum_{j=1}^{m_0}\int_0^t K_j(t-s)v_j(s)ds$$

$$+\Phi_\nu(A-I)^{-1}(A-I)^{-1}aN_1(v,\bar{K}\bar{C}\bar{v}\bar{w}) + f_{1\nu}(t)\Big\},$$

$$+C_j(t) = \sum_{\nu=1}^{m} \beta_{m_0+j,\nu}(t)\Big\{ -\Phi_\nu(A-I)^{-1}\int_0^t \sum_{j=1}^{m_0} K_j(t-s)v_j(s)ds$$

$$\Phi_\nu(A-I)^{-1}(A-I)^{-1}aN_1(v,\bar{K}\bar{C}\bar{v}\bar{w}) + f_{1,\nu}(t)\Big\}, \qquad (3.3.31)$$

where $i = 1,...,m_0$; $j = 1,...,m_1$; $\bar{u} = (u_1,...,u_{m_0})$; $\bar{w} = (w_1,...,w_{m_1})^t$; $\bar{K} = (K_1...,K_{m_0})$; $\bar{C} = (C_1,...,C_{m_1})$; $\bar{v} = (v_1,...,v_{m_0})^t$, $f_1(t)$ and $f_2(t)$ are defined by formulae (3.3.13) and (3.3.17) respectively, and

$$N_0(u,\bar{K},\bar{C},\bar{u},\bar{w}) = \int_0^t e^{-a(t-s)}\Bigg[au(s) + \sum_{j=1}^{m_0}\int_0^s K_j(s-\tau)u_j(\tau)d\tau$$

$$+\sum_{j=1}^{m_0}\int_0^s C_j(\tau)w_j(\tau)d\tau\Bigg]\, ds. \qquad (3.3.32)$$

$$N_1(v,\bar{K},\bar{C},\bar{v},\bar{w}) = \int_0^t e^{-a(t-s)}[av(s) + \sum_{i=1}^{m_0} K_i(s)B_i u_0$$

$$+\sum_{i=1}^{m_0}\int_0^s K_i(s-\tau)v_i(\tau)d\tau + \sum_{i=1}^{m_1} C_i(s)w_i(s)]ds. \qquad (3.3.33)$$

From the condition $D(A) \subset D(B_i)$, $D(A) \subset D(E_j)$, $i = 1,...,m_0$, $j+1,...,m_1$ it follows that the operators $A(A-I)^{-1}$, $B_i(A-I)^{-1}$, and $E_j(A-I)^{-1}$ are the elements from $L(X)$.

Introduce the space

$$Z = Z_1 \times Z_2 \times Z_3 \times Z_4 \times Z_5 \times Z_6 \times Z_7$$

where

$$Z_1 = Z_2 = C([0,T];X);\ Z_3 = Z_4 = C_{m_0}([0,T];X)$$

$$Z_5 = C_{m_1}([0,T];X);\ Z_6 = C_{m_0}[0,T];\ Z_7 = C_{m_1}[0,T]$$

Here, $C_m[0,T]$ is the space of m-dimensional vector-functions with the coordinates from $C[0,T]$. Define in $C_m[0,T]$ the norms

$$\|a(t)\|_{C_m[0,T]} = \sup_{t\in[0,T]} \left\{ \max_i |a_i(t| \right\},$$

where $a(t) = (a_1, ..., a_m)^t$.

Introduce the notations

$$\|\cdot\|_1 = \|\cdot\|_2 = \|\cdot\|_C,\ \|\cdot\|_3 = \|\cdot\|_4 = \|\cdot\|_{C,m_0},$$

$$\|\cdot\|_5 = \|\cdot\|_{Cm_1},$$

$$\|\cdot\|_5 = \|\cdot\|_{Cm_1},\ \|\cdot\|_6 = \|\cdot\|_{C_{m_0}[0,T]},\ \|\cdot\|_7 = \|\cdot\|_{C_{m_1}[0,T]},$$

$$\|\cdot\|_Z = \sum_{i=1}^{7} \|\cdot\|_i,\ \|\cdot\|_i = \|\cdot\|_{Z_i},\ i = 1, ..., 7.$$

We shall show the existence of unique fixed points of the mapping

$$S : Z \to Z,$$

$$S : Z = \begin{pmatrix} u \\ v \\ \bar{u} \\ \bar{v} \\ w \\ \bar{K} \\ \bar{C} \end{pmatrix} \to Sz = \begin{pmatrix} S_1 z \\ S_2 z \\ S_3 z \\ S_4 z \\ S_5 z \\ S_6 z \end{pmatrix}, \qquad (3.3.34)$$

where

$$S_1 z = -(A-I)^{-1} N_0(u, \bar{u}, \bar{K}, \bar{C}, \bar{w}) + (A-I)^{-1} f_1(t),$$

$$S_2 z = -(A-I)^{-1} N_1(v, \bar{K}, \bar{C}, \bar{v}, \bar{w}) + (A-I)^{-1} f_2(t),$$

$$(S_3 z)_i = -B_i(A-I)^{-1} N_0(u, \bar{K}, \bar{C}, \bar{u}, \bar{w}) + B_i(A-I)^{-1} f_1(t),$$

$$(S_4 z)_i = -B_i(A-I)^{-1} N_1(v, \bar{K}, \bar{C}, \bar{v}, \bar{w}) + B_i(A-I)^{-1} f_2(t),$$

$$(S_5 z)_j = -E_j(A-I)^{-1} N_0(u, \bar{K}, \bar{C}, \bar{u}, \bar{w}) + E_j(A-I)^{-1} f_1(t),$$

$$(S_6 z)_i = \sum_{\nu=1}^{m} \beta_{i\nu}(t)\Big\{ - \Phi_\nu (A-I)^{-1}\Big[\sum_{j=1}^{m_0} \int_0^t K_j(t-s) v_j(s) ds\Big]$$

$$+\Phi_\nu (A-I)^{-1}(A-I)^{-1} a N_1(v, \bar{K}, \bar{C}, \bar{v}, \bar{w}) + f_{1,\nu}(t)\Big\}$$

$$(S_7 z)_i = \sum_{\nu=1}^{m} \beta_{m_0+j,\nu}(t)\Big\{ - \Phi_\nu (A-I)^{-1}\Big[\sum_{j=1}^{m_0} \int_0^t K_j(t-s) v_j(s) ds\Big]$$

$$+\Phi_\nu (A-I)^{-1}(A-I)^{-1} a N_1(v, \bar{K}, \bar{C}, \bar{v}, \bar{w}) + f_{1,\nu}(t)\Big\}$$

where $i = 1, ..., m_)$; $j = 1, ...m_1$; $m = m_0 + m_1$.

Formulae (3.3.13), (3.3.17), and (3.3.27) yield that there exists such R that

$$\|(A-I)^{-1} f_1(t)\|_1 + \|(A-I)^{-1} f_2(t)\|_2$$

$$+\sum_{j=1}^{2} \sup_{t\in[0,T]} \left\{ \max_{1\le i\le m_0} \|B_i(A-I)^{-1} f_j(t)\|_X \right\}$$

$$+ \sup_{t\in[0,T]} \left\{ \max_{1\le j\le m_1} \|E_j(A-I)^{-1} f_1(t)\|_X \right\}$$

$$+ \sup_{t\in[0,T]} \left\{ \max_{1\le i\le m_0} \left| \sum_{\nu=1}^{m} \beta_{1\nu}(t) f_{1\nu}(t) \right| \right\}$$

$$+ \sup_{t\in[0,T]} \left\{ \max_{1\le j\le m_1} \left| \sum_{\nu=1}^{m} \beta_{m_0\nu}(t) f_{1\nu}(t) \right| \right\} \le R \qquad (3.3.35)$$

Consider the ball

$$Z(2R) = \{(u, v, u^{-t}, v^{-t}, w^{-t}, K^{-t}, C^{-t})^t \in Z :$$

$$\|u\|_1 + \|v\|_2 + \|u\|_3 + \|v\|_4 + \|w\|_5 + \|\bar{K}\|_6 + \|\bar{C}_7\| \le 2R\} \qquad (3.3.36)$$

We shall show that for sufficiently small T the mapping S maps the ball $Z(2R)$ into intself and will be contractive.

For each $T' \in (0, T]$ from (3.3.32) and (3.3.33) we have

$$\|N_0(u, \bar{K}, \bar{C}, \bar{u}, \bar{w})\|_X \le NT'[\|u(t)\|_1 + \|\bar{u}\|_3 + \|u\|_3\|\bar{K}\|_6$$

$$+ \|\bar{w}\|_5\|\bar{C}\|_7], \qquad (3.3.37)$$

$$\|N_1(v, \bar{K}\bar{C}\bar{v}\bar{w})\|_X \le NT'[\|v(t)\|_2 + \|\bar{K}\|_6(1$$

$$+ \|\bar{v}\|_4) + \|\bar{w}\|_5\|\bar{C}\|_7], \qquad (3.3.38)$$

where $N \ge 0$ is a constant.

Taking into account (3.3.35), (3.3.36), (3.3.37) and (3.3.38) from (3.3.34) we obtain

$$\|Sz\|_Z \le R + M_0(R+1)RT' \qquad (3.3.39)$$

where $M_0 > 0$ is a constant not depending on R or T'.

From (3.3.39) if follows that

$$\|Sz\|_Z \le 2R, \tag{3.3.40}$$

for sufficiently small $t' \in (0, T]$.

Now, let

$$z_1 = (u_1, v_1, \bar{u}_1^t, \bar{v}_1^t, \bar{w}_1^t, \bar{K}_1^t \bar{C}_1^t)^t$$

$$z_2 = (u_2, v_2, \bar{u}_2^t, \bar{v}_2^t, \bar{w}_2^t, \bar{K}_2^t \bar{C}_2^t)^t \in Z(2R).$$

Introduce the notations

$$x = N_0(u_1, \bar{K_1}, \bar{C_1}, \bar{u_1}, \bar{w_1}) - N_0(u_2, \bar{K_2}, \bar{C_2}, \bar{u_2}, \bar{w_2}), \tag{3.3.41}$$

$$y = N_1(v_1, \bar{K_1}, \bar{C_1}, \bar{v_1}, \bar{w_1}) - N_1(v_2, \bar{K_2}, \bar{C_2}, \bar{v_2}, \bar{w_2}) \tag{3.3.42}$$

Then

$$S_1 z_1 - S_1 z_2 = -(A - I)^{-1} x,$$

$$S_2 z_1 - S_2 z_2 = -(A - I)^{-1} y,$$

$$(S_3 z_1)_i - (S_3 z_2)_i = -B_i (A - I)^{-1} x, \tag{3.3.43}$$

$$(S_4 z_1)_i - (S_4 z_2)_i = -B_i (A - I)^{-1} y,$$

$$(S_5 z_1)_j - (S_5 z_2)_j = -E_j (A - I)^{-1} x,$$

$$(S_6 z_1)_i - (S_6 z_2)_i = \sum_{\nu=1}^{m} \beta_{1,\nu}(t)$$

$$\Big\{ - \Phi_\nu (A - I)^{-1} \sum_{j=1}^{m_0} \int_0^t \Big[K_j^1(t - s) v_j^1(s) - K_j^2(t - s) v_j^2(s) ds \Big]$$

$$+ a \Phi_\nu (A - I)^{-1} (A - I)^{-1} y \Big\},$$

$$(S_7z_1)_j - (S_7z_2)_j = \sum_{\nu=1}^{m} \beta_{m_0+j,\nu}(t)$$

$$\Big\{ - \Phi_\nu(A-I)^{-1}\sum_{i=1}^{m_0}\int_0^t \Big[K_i^1(t-s)v_i^1(s) - K_i^2(t-s)v_i^2(s)ds\Big]$$

$$+a\Phi_\nu(A-I)^{-1}(A-I)^{-1}y\Big\}$$

where $i = 1,\dots, m_0$; $j = 1,\dots, m_1$. $\bar{K}_p = (K_1^p,\dots, K_{m_0}^p)$, $\bar{v}_p = (v_1^p,\dots,v_{m_0}^p)$, $p = 1,2$.

Taking into account (3.3.32) and (3.3.33) for each $T' \in (0,T]$ from (3.3.41) and (3.3.42), we obtain

$$\|x\|_X \le M_0T'\{\|u_1-u_2\| + 2R[\|\bar{u}_1-\bar{u}_2\|_3$$

$$+\|\bar{K}_1-\bar{K}_2\|_6 + \|\bar{w}_1-\bar{w}_2\|_5 + \|\bar{C}_1-\bar{C}_2\|_7]\}, \tag{3.3.44}$$

$$\|y\|_X \le M_0T'\{\|v_1-v_2\|_2 + \|\bar{K}_1-\bar{K}_2\|_6 + 2R[\|\bar{v}_1-\bar{v}_2\|_4$$

$$+\|\bar{K}_1-\bar{K}_2\|_6 + \|\bar{w}_1-\bar{w}_2\|_5 + \|\bar{C}_1-\bar{C}_2\|_7]\}, \tag{3.3.45}$$

where $M_0 > 0$ is a constant.

The conditions I-IV, estimates (3.3.44), (3.3.45), and (3.3.43) yield

$$\|Sz_1 - Sz_2\|_Z \le M_1T'(R+1)\|z_1-z_2\|_Z, \tag{3.3.46}$$

where $M_1 > 0$ is a constant not depending on R or T'.

From (3.3.39) and (3.3.46) it follows for sufficiently small $T' \in (0,T]$ the operator s maps the ball $Z(2R)$ into intself and will be contractive. Thus, the following theorem is established

Theorem 3.3.1. Let the above conditions hold. Then such $t' \in (0,T]$ exists that inverse problem (3.3.1) - (3.3.3) has a unique solution $\{u(t), K_1(t),\dots,C_{m_1}(t)\}$ from the space

$$C^1([0,T'];\ D(A))\cdot C_{m_0}[0,T']\cdot C_{m_1}[0,T'].$$

The stability estimate. Let $M(t)$ be $m \cdot m$ a matrix - valued function from the condition IV.

Definition. The matrix-valued function $M(t) \in K(\beta)$ if and only if the inverse $M^{-1}(t)$ exists and

$$\sup_{t\in[0,T]} \|M^{-1}(t)\| \leq \beta,$$

where $\beta > 0$ is a constant, $\|\cdot\|$ is the matrix norm.

Let $\{u_l(t), \bar{K}_l(t), \bar{C}_l(t)\}, (l = 1,2)$ be two solutions to inverse problem (3.3.1) - (3.3.3) from the space $C^1([0,T]; D(A)) \cdot C_{m_0}[0,T] \cdot C_{m_1}[0,T]$ with the data $f_l(t)$, $u_{0l}, g_{il}(t)$, $i = 1, \dots, m$, where

$$\bar{C}_l(t) = \{C_{1,l}(t), \dots, C_{m_1,l}(t)\}^t,$$

$$\bar{K}_l(t) = \{K_{1,l}(t), \dots, K_{m_0,l}(t)\}^t,$$

Then

$$u'_{lt}(t) = (Au_l)'_t + a(Au_l) + \sum_{i=1}^{m_0} \int_0^t K_{i,l}(t-s) \tag{3.3.47}$$

$$B_i u_e(s) ds + \sum_{i=1}^{m_1} \int_0^t C_{i,l}(s) E_i u_e(s) ds + f_e(t) \tag{3.3.48}$$

$$u_e(0) = u_{0e},\ u_{0e} \in D(A)$$

$$\Phi_i u_e(t) = g_{ie}(t),\ t \in [0,T], i = 1, \dots, m \tag{3.3.49}$$

Further we assume that

V. $M_e(t) \in K(\beta)$, $l = 1,2$, where $M_e(t)$ is the following matrix (denote, for bravity, by A^* the operator $(A-I)^{-1}$)

$$\begin{pmatrix} \Phi_1 A^* B_1 u_{0l} & \dots & \Phi_1 A^* Bm_0 u_{0l} \alpha_{11} g_{1l}(t) & \dots & \alpha_{1m_1} g_{1l}(t) \\ \Phi_2 A^* B_1 u_{0l} & \dots & \Phi_2 A^* Bm_0 u_{0l} \alpha_{21} g_{2l}(t) & \dots & \alpha_{2m_1} g_{2l}(t) \\ \dots\dots & \dots\dots\dots & \dots\dots\dots & \dots\dots & \dots\dots \\ \Phi_m A^* B_1 u_{0l} & \dots & \Phi_m A^* Bm_0 u_{0l} \alpha_{m1} g_{ml}(t) & \dots & \alpha_{mm_1} g_{ml}(t) \end{pmatrix}.$$

Introduce the notations (A)

$$u_3(t) = u_1(t) - u_2(t),\ t \in [0, T].$$

$$\bar{K}_3(t) = \bar{K}_1(t) - \bar{K}_2(t), \quad \bar{C}_3(t) = \bar{C}_1(t) - \bar{C}_2(t),$$

$$u_{03} = u_{01} - u_{02},\ f_3(t) = f_1(t) - f_2(t),$$

$$g_{i3} = g_{i1}(t) - g_{i2}(t), i = 1, \ldots, m.$$

$$\Phi = \begin{pmatrix} \Phi_1 \\ \vdots \\ \Phi_m \end{pmatrix},\ g_k(t) = \begin{bmatrix} g_{1k}(t) \\ \vdots \\ g_{mk}(t) \end{bmatrix}, k = 1, 2, 3,$$

$$\bar{u}_k(t) = \begin{bmatrix} u_{1k}(t) \\ \vdots \\ u_{m_0 k}(t) \end{bmatrix},\ u_{1,k}(t) = B_i u_k(t), i = 1, \ldots, m_0,$$

$$v_k(t) = u'_k(t),\ \bar{v}_k(t) = \begin{bmatrix} v_{1k}(t) \\ \vdots \\ u_{m_0 k}(t) \end{bmatrix},\ v_{1,k}(t) = B_i v_k(t),$$

$$\bar{w}_k(t) = \begin{bmatrix} w_{1k}(t) \\ \vdots \\ w_{m_1 k}(t) \end{bmatrix},\ w_{j,k}(t) = E_j u_k(t), j = 1, \ldots, m_1.$$

Then, from (3.3.47) we have

$$u'_3(t) = -(Au_3)'^{1_t} + a(Au_3) + \sum_{i=1}^{m_p} \int_0^t \Big[K_{i1}(t-s) B_i u_3(s)$$

$$+ K_{i31}(t-s) B_i u_2(s)\Big] ds + \sum_{i=1}^{m_0} \int_0^t \Big[C_{i1}(s) E_i u_3(s)$$

$$+ C_{i3}(s) E_i u_2(s)\Big] ds + f_3(t), \tag{3.3.50}$$

$$u_3(0) = u_{03} \in D(A),$$

$$\Phi_i u_3(t) = g_{i3}(t),\ i = 1, \dots, m. \tag{3.3.51}$$

With an analogous procedure and taking into account the notations (A), from (3.3.49) - (3.3.50) we obtain the following system

$$u_3(t) = -(A-I)^{-1}\int\limits_0^t l^{-a(t-s)}\Big\{au_3(s)$$

$$+\sum_{i=1}^{m_0}\int\limits_0^s\Big[K_{i1}(s-\tau)u_{i3}(\tau)+K_{i3}(s-\tau)u_{i2}(\tau)\Big]d\tau$$

$$+\sum_{j=1}^{m_1}\int_0^s\Big[C_{j1}(\tau)w_{j3}(\tau)+C_{j3}(\tau)w_{j2}(\tau)\Big]d\tau ds+F_1(t), \tag{3.3.52}$$

$$v_3 = -(A-I)^{-1}\int\limits_0^t l^{-a(t-s)}\Big\{av_3(s)$$

$$+\sum_{i=1}^{m_0}\Big[K_{i1}(s)u_{i3}(0)+K_{i3}(s)u_{i2}(0)\Big]+\sum_{j=1}^{m_1}\Big[C_{j1}(s)w_{j3}(s)$$

$$+C_{j3}(s)w_{j2}(s)\Big]ds+\sum_{i=1}^{m_0}\int\limits_0^s\Big[K_{i1}(s-\tau)v_{is}(\tau)$$

$$+K_{i3}(s-\tau)v_{i2}(\tau)\Big]d\tau\Big\}ds+F_2(t), \tag{3.3.53}$$

$$u_{i3}(t) = -B_i(A-I)^{-1}\int\limits_0^t l^{-a(t-s)}\Big\{au_3(s)$$

$$+\sum_{i=1}^{m_0}\int\limits_0^s\Big[K_{i1}(s-\tau)u_{i3}(s)+K_{i3}(s-\tau)u_{i2}(\tau)\Big]d\tau$$

$$+\sum_{j=1}^{m_1}\int_0^s\Big[C_{j1}(\tau)w_{j3}(\tau)+C_{j3}(\tau)w_{j2}(\tau)\Big]d\tau\Big\}ds+N_iF_1(t), \tag{3.3.54}$$

$$w_{j3}(t) = -E(A-I)^{-1}\int_0^t l^{-a(t-s)}\Big\{au_3(s)$$

$$+\sum_{i=1}^{m_0}\int_0^s\Big[K_{i1}(s-\tau)u_{i3}(\tau)+K_{i3}(s-\tau)u_{i2}(\tau)\Big]d\tau$$

$$+\sum_{j=1}^{m_1}\int_0^s\Big[C_{j1}(\tau)w_{j3}(\tau)+C_{j3}(\tau)w_{j2}(\tau)\Big]d\tau\}ds + F_jF_1(t), \quad (3.3.55)$$

$$v_{i3}(t) = -B_i(A-I)^{-1}\int_0^t l^{-a(t-s)}\Big\{av_3(s)$$

$$+\sum_{i=1}^{m_0}\Big[K_{i1}(s)u_{i3}(0)+K_{i3}(s)u_{i2}(0)\Big]$$

$$+\sum_{j=1}^{m_1}\Big[C_{j1}(s)w_{j3}(s)+C_{j3}(s)w_{j2}(s)\Big]\sum_{i=1}^{m_0}\int_0^s\Big[K_{i1}(s-\tau)v_{i3}(\tau)$$

$$+K_{i3}(s-\tau)v_{i2}(\tau)\Big]d\tau\}ds + B_iF_2(t), \quad (3.3.56)$$

where $i = 1,\dots,m_0$; $j = 1,\dots m_1$,

$$F_1(t) = (A-I)^{-1}\Big[l^{at}(Au_{03}-u_{03}) - \int_0^t l^{-a(t-s)}f_3(s)ds\Big] \quad (3.3.57)$$

$$F_2(t) = -(A-I)^{-1}\Big[l^{-at}(aAu_{03}+f_0(0)) + \int_0^t l^{-a(t-s)}f_3'(s)ds\Big]. \quad (3.3.58)$$

Differentiating (3.3.49) by t we have

$$(B_ju_{02})K_{j\sum_{j=1}^{m_0}3}(t) + \sum_{j=1}^{m1}C_{j3}(t)E_ju_2(t)$$

$$= -\sum_{j=1}^{m_0}(B_ju_{03})K_{j1}(t) - \sum_{j=1}^{m1}C_{j1}(t)E_ju_3(t)$$

$$-\sum_{j=1}^{m_0}\int_0^t [K_{j3}(t-s)B_jv_2(s)+K_{j1}(t-s)B_jv_3(s)]ds$$

$$-[(A-I)v_3(t)]'_t - a(A-I)v_3(t) - av_3(t) - f'_3(t). \qquad (3.3.59)$$

Operating by $M_2^{-1}(t)\Phi(A-I)^{-1}$ from the left on equation (3.3.58), and using notations (A) we shall have

$$\begin{pmatrix} \bar{K}_3 & (t) \\ \bar{C} & (t) \end{pmatrix} = -M_2^{-1}(l)\Phi(A-I)^{-1}\Big\{\sum_{j=1}^{m_0} C_{j1}(t)w_{j3}(t)$$

$$+\sum_{j=1}^{m_0}\int_0^t K_{j3}(t-s)v_{j2}(s)+K_{j1}(t-s)v_{j3}(s)]ds+av_3(t)\Big\}+F_3(t), \quad (3.3.60)$$

where $M_2^{-1}(t)$ is the inverse to $M_2(t)$, and

$$F_3(t) = -M_2^{-1}(t)[g''(t)+ag'_3(t)]$$

$$-M_2^{-1}(t)\Phi(A-I)^{-1}f'_3(t)$$

$$-M_2^{-1}(t)\Phi(A-I)^{-1}\sum_{j=1}^{m_0}(B_ju_{03})K_{j1}(t). \qquad (3.3.61)$$

Introduce the following notations

$$\Psi = \|f_3(t)\|_{C^1(0,T;X)} + \|u_03\|_X + \|Au_03\|_X$$

$$+\|g'_3\|_{C_m[0,T]} + \|g''_3(t)\|_{C_m[0,T]} + \sum_{i=1}^{m_0}\|u_{i3}(t)\|_X \qquad (3.3.62)$$

$$\xi(t) = \|u_3(t)\|_X + \|v_3(t)\|_X + \sum_{i=1}^{m_0}\|u_{i3}(t)\|_X$$

$$+\sum_{i=1}^{m_0}\|v_{i3}(t)\|_X + \sum_{j=1}^{m_1}\|w_{j3}(t)\|_X + \sum_{i=1}^{m_0}|K_{i3}(t)|$$

$$+\sum_{j=1}^{m_1}|C_{j3}(t)|,\ t\in[0,T] \tag{3.3.63}$$

For the positive fixed R_0, R_1, and R_2 determine

$$Q_1(R_0)=\Big\{u\in C^1([0,T];D(A)):\|u(t)\|_{C^1([0,T];X)}$$

$$+\sum_{i=1}^{m_0}\|B_iu(t)\|_{C^1(0,T);X}+\sum_{j=1}^{m_1}\|E_ju(T)\|_{C([0,T];X)}\le R_0\Big\} \tag{3.3.64}$$

$$Q_2(R_1)=\Big\{\bar{K}(t)\in C_{m_0}([0,T]:\|\bar{K}(t)\|_{C_{m_0}[0,T]}\le R_1\Big\} \tag{3.3.65}$$

$$Q_3(R_2)=\Big\{\bar{C}(t)\in C_{m_1}([0,T]:\|\bar{C}(t)\|_{C_{m_1}[0,T]}\le R_2\Big\} \tag{3.3.66}$$

Further, we shall assume that

$$u_1(t),u_2(t)\in Q_1(R_0),\quad \bar{K}_1(t),\bar{K}_2(t)\in Q_2(R_1),$$

$$\bar{C}_1(t),\bar{C}_2(t)\in Q_3(R_2). \tag{3.3.67}$$

Taking into account the formulae (3.3.56), (3.3.57), (3.3.60), (3.3.62), (3.3.63), (3.3.53), (3.3.54), (3.3.55), and (3.3.59) we obtain

$$\xi(t)\le q_1\int_0^t(s)ds+q_0\Psi,t\in[0,T], \tag{3.3.68}$$

where $q=q_0\exp(q,T)$.

Taking into account the notations (A), (3.3.61), and (3.3.62), from (3.3.68) we obtain

$$\|u_1(t) - u_2(t)\|_{C^1([0,T];X)} + \sum_{i=1}^{m_0} \|B_i[u_1(t) - u_2(t)]\|_{C^1([0,T];X)}$$

$$+ \sum_{i=1}^{m_1} \|E_i[u_1(t) - u_2(t)]\|_{C([0,T];X)}$$

$$+\|\bar{K}_1(t) - \bar{K}_2(t)\|_{C_{m_0}[0,T]} + \|\bar{C}_1(t) - \bar{C}_2(t)\|_{C_{m_1}[0,T]}$$

$$\leq q[\|f_1(t) - f_2(t)\|_{C^1([0,T];X)}$$

$$+\|Au_{01} - Au_{02}\|_X + \sum_{i=1}^{m_0} \|B_i(u_{01} - u_{02})\|_X$$

$$+ |g_1'(t) - g_2'(t)\|_{C_m[0,T]} + \|g_1''(t) - g_2''(t)\|_{C_m[0,T]} \tag{3.3.69}$$

From (3.3.51) we have

$$Au_3(t) = -A(A-I)^{-1} \int_0^t l^{-a(t-s)}\{au_3(s)$$

$$+\sum_{i=1}^{m_0} \int_0^s [K_{i1}(s-\tau)u_{i3}(\tau) + K_{i3}(s-\tau)u_{i2}(\tau)]d\tau$$

$$+\sum_{j=1}^{m_1} \int_0^s [C_{j1}(\tau)w_{j3}(\tau) + C_{j3}(\tau)w_{j2}(\tau)]d\tau\Big\}ds + AF_1(t), \tag{3.3.70}$$

By condition (3.3.66), estimate (3.3.69), and notations (A), from (3.3.51) we obtain

$$\|Au_1(t) - Au_2(t)\|_{C([0,T];X)} \leq q_2\Psi, \tag{3.3.71}$$

where $q_3 > 0$ is a constant.

Analogously, from (3.3.53) we have

$$\|Av_1(t) - Av_2(t)\|_{C([0,T];X)} \leq q_3\Psi, \tag{3.3.72}$$

where $g_3 > 0$ is a constant. Estimates (3.3.69), (3.3.71), and (3.3.72) yield

$$\|u_1(t) - u_2(t)\|_{C^1([0,T];X)} + \|Au_1(t) - Au_2(t)\|_{C^1([0,T];X)}$$

$$+\sum_{i=1}^{m_0} \|B_i u_1(t) - B_i u_2(t)\|_{C^1([0,T];X)}$$

$$+\sum_{i=1}^{m_1} \|E_i u_1(t) - E_i u_2(t)\|_{C([0,T];X)}$$

$$+\sum_{i=1}^{m_1} \|E_i[u_1(t) - u_2(t)]\|_{C([0,T];X)}$$

$$+\|\bar{K}_1(t) - \bar{K}_2(t)\|_{C_{m_0}[0,T]} + \|\bar{C}_1(t) - \bar{C}_2(t)\|_{C_{m_1}[0,T]}$$

$$\leq q[\|f_1(t) - f_2(t)\|_{C^1([0,T];X)}$$

$$+\|Au_{01} - Au_{02}\|_X + \sum_{i=1}^{m_0} \|B_i(u_{01} - u_{02})\|_X$$

$$+ \|g_1'(t) - g_2'(t)\|_{C_m[0,T]} + \|g_1''(t) - g_2''(t)\|_{C_m[0,T]}. \qquad (3.3.73)$$

where $g_n > 0$ is a constant depending on R_0, R_1, R_2, a and on

$$T, \beta, \|(A-I)^{-1}\|_{L(X)}, \quad \|A(A-I)^{-1}\|_{L(X)},$$

$$\|B_i(A-I)^{-1}\|_{L(X)}, \quad \|E_j(A-I)^{-1}\|_{L(X)},$$

$$\|\Phi\|_{L(X,R)}, \quad (i = 1, \ldots .m_0; j = 1, \ldots, m_1; k = 1, \ldots, m).$$

Thus, we have solved the following theorem:

Theorem 3.3.2. Let the condition IV hold, and

$$\begin{pmatrix} u_1 & (t) \\ \bar{K}_1 & (t) \\ \bar{C}_1 & (t) \end{pmatrix} \text{ and } \begin{pmatrix} u_2 & (t) \\ \bar{K}_2 & (t) \\ \bar{C}_2 & (t) \end{pmatrix}$$

be two solutions to inverse problem (3.3.1) - (3.3.3) from the class $Q_1(R_0) \cdot Q_2(R_1) \cdot Q_3(R_2)$ with data $f_l(t)$, u_{0l}, $g_l(t), l = 1, 2$. Here $Q(R_0), Q_2(R_1)$ and $Q_3(R_2)$ are defined by formulae (3.3.63), (3.3.64), and (3.3.65),

$$f_1(t), f_2(t) \in C^1([0,T]; X),$$

$$u_{01}, u_{02} \in D(A), g_1''(t), g_2''(t) \in C_m[0,T].$$

Then the stability estimate (3.3.73) holds.

As an example, consider the following inverse problem

$$\frac{\partial u}{\partial t} = \frac{\partial}{\partial t}(\frac{\partial^2 u}{\partial x^2} + w^2 u) + a(\frac{\partial^2 u}{\partial x^2} + w^2 u)$$

$$+\sum_{i=1}^{m_0} \int_0^t K_i(t-s)\Big[b_{i1}(x)\frac{\partial^2 u(s,x)}{\partial x^2} + b_{i2}(x)\frac{\partial u(s,x)}{\partial x}$$

$$+b_{i3}(x)u(s,x)\Big]ds + \sum_{i=1}^{m_1} \int_0^t C_i(s)\Big[\frac{-\partial^2(d_i(x)u(s,x))}{\partial x^2}\frac{\partial u(s,x)}{\partial x}$$

$$+ (w^2-1)d_i(x)u(s,x)\Big]ds + f(t,x). \qquad (3.3.74)$$

$$u(0,x) = u_0(x) \qquad 0 \le x \le 1, \qquad (3.3.75)$$

$$u(t,0) = u_(t,1) = 0, \qquad 0 \le t \le T, \qquad (3.3.76)$$

$$u(t,x_i) = g_i(t), \qquad 0 \le t \le T,\ i = 1,\ldots,m \qquad (3.3.77)$$

where $m = m_0 + m_1$, $0 \leq x_1 \leq x_2 \leq \ldots \leq x_m \leq 1$. a, w are given numbers $b_{i1}(x)$, $b_{i2}(x)$, $b_{i3}(x)$ $i = 1, 2\ldots, m_0$ are given continuous functions from $[0,1]$. $f(t,x)$ is given and belongs to $C^1([0,T])$; C([0,1]). $u_0(x)$, $d_j(x)$, $j = 1,\ldots,m_1$ are given and from $C^2[0,1]$. $g_i(t)$, $i = \ldots, m$ are given and from $C^2[0,T]$. The functions $K_i(t)$, $C_j(t)$, $i = 1,\ldots m_0$; $j = 1,\ldots,m_1$ are unknowns from $C[0,T]$. $u(t,x)$ is an unknown function from $C^1([0,T]; C^2[0,1])$.

Here $D(A) = \{v(x) \in C^2[0,1]; v(0) = v(1) = 0\Big\}$,

$$A = \frac{\partial^2}{\partial x^2} + w^2, B_i = b_{i1}(x)\frac{\partial^2}{\partial x^2} + b_{i2}\frac{\partial}{\partial x} + b_{i3}(x),$$

$$E_j = \frac{\partial^2}{\partial x^2}[d_j(x)\cdot] + (w^2 - 1)d_j(x), i = 1,\ldots,m_0;$$

$j = 1,\ldots,m_1 X = C[0,1]$, I is the unit operator.

$$\Phi_i v(x) = v(x_i), i + 1,\ldots,m;$$

$$v(x) \in X, \|v(x)\|_X = \sup_{x\in[0,1]} |v(t)|.$$

Assume everywhere that $w^2 > 1$, $\sqrt{w^2 - 1} \not\equiv \pi n$, $n \in N$, the set of natural numbers. Then, for inverse problem (3.3.74) - (3.3.77) the condition I-III hold. It is well known that

$$(A - I)^{-1}v(x) = -\int_0^x \frac{\sin(\sqrt{w^2 - 1}(1 - x))\sin\sqrt{w^2 - 1}\xi}{\sqrt{w^2 - 1}\sin\sqrt{w^2 - 1}} v(\xi)d\xi$$

$$-\int_1^x \frac{\sin(\sqrt{w^2 - 1}x)\sin(\sqrt{w^2 - 1}(1 - \xi))}{\sqrt{w^2 - 1}\sin\sqrt{w^2 - 1}} v(\xi)d\xi. \quad (3.3.78)$$

Introduce the notation

$$a_{ij} = \Phi_i(A - I)^{-1}B_j u_0(x). \quad i = 1,\ldots,m;\ j = 1,\ldots,m_1. \quad (3.3.79)$$

Then, by formula (3.3.78), from (3.3.79) we have

$$a_{ij} = -\int_0^{x_i} \frac{\sin(\sqrt{w^2-1}(1-x_i))\sin(\sqrt{w^2-1}\xi)}{\sqrt{w^2-1}\sin\sqrt{w^2-1}} \Big[b_{j1}(\xi)$$

$$+b_{j2}(\xi)u_0'(\xi) + b_{j3}(\xi)u_0(\xi)\Big]d\xi$$

$$-\int_{x_i}^{1} \frac{\sin(\sqrt{w^2-1}x_i)\sin(\sqrt{w^2-1}(1-\xi))}{\sqrt{w^2-1}\sin\sqrt{w^2-1}} [b_{j1}(\xi)u_0''(\xi)$$

$$+ b_{j2}(\xi)u_0'(\xi) + b_{j3}(\xi)u_0(\xi)]d\xi, \quad (i = 1,\ldots,m. \quad j = 1,\ldots,m_1). \tag{3.3.80}$$

For each $v(x) \in D(A)$, due to (3.3.78) we obtain

$$(A-I)^{-1}E_jv(x) = -\int_0^{x} \frac{\sin(\sqrt{w^2-1}(1-x))\sin\sqrt{w^2-1}\xi}{\sqrt{w^2-1}\sin\sqrt{w^2-1}}$$

$$\times\Big\{\frac{d^2}{d\xi^2}[d_j(\xi)v(\xi)] + (w^2-1)d_j(\xi)v(\xi)d\xi$$

$$-\int_x^1 \frac{\sin(\sqrt{w^2-1}x)\sin(\sqrt{w^2-1}(1-\xi))}{\sqrt{w^2-1}\sin\sqrt{w^2-1}} \Big\{\frac{d^2}{d\xi^2}[d_j(\xi)v(\xi)\Big]$$

$$+(w^2-1)d_j(\xi)v(\xi)\Big\}d\xi = d_j(x)v(x). \quad j = 1,\ldots,m_1$$

Therfore, for each $v(x) \in D(A)$ we have

$$\Phi_i(A-I)^{-1}E_jv(x) = d_j(x_i)v(x_i), \tag{3.3.81}$$

where $i = 1,\ldots,m$; $j = 1,\ldots,m_1$.

Taking into account (3.3.74), (3.3.79), and (3.3.81), the matrix $M(t)$ may be represented as follows

$$\begin{pmatrix} a_{11} & a_{12} & \ldots & a_{1m_0} & d_1(x_1) & g_1(t) & \ldots & d_{m_1}(x_1) & g_1(t) \\ a_{21} & a_{22} & \ldots & a_{2m_0} & d_1(x_2) & g_2(t) & \ldots & d_{m_1}(x_2) & g_2(t) \\ \ldots & \ldots & \ldots & \ldots & \ldots & \ldots & \ldots & \ldots & \ldots \\ a_{m1} & a_{m2} & \ldots & a_{mm_0} & d_1(x_m) & g_m(t) & \ldots & d_{m_1}(x_m) & g_m(t) \end{pmatrix}$$

where a_{ij} are determined from (3.3.79), $i = 1, \ldots, m$; $j = 1, \ldots, m_0$

Assume that

$$\det M(t) \neq 0, \tag{3.3.82}$$

for $t \in [0, T]$. Condition (3.3.83) implies the condition IV. Therefore, Theorems 3.3.1, 3.3.2 hold for inverse problem (3.3.74)– (3.3.77) if the condition V holds.

Bibliography

Abdukarimov, A.(1983). *Uniqueness and Stability of Problems on Continuation of Solution of Elliptic and Parabolic Equations from Discrete Sets.* Dissert. Kandid. fiz. - mat. nauk. Novosibirsk (in Russian).

Amonov, B. K. and Shishatskii, S.P. (1972). *A priori* estimate of the Cauchy problem solution with data in the time-like surface for the parabolic equation of the second order and the uniqueness theorem connected with it. *Dokl. Akad. Nauk SSSR* **206**, (1), 11-12 (in Russian).

Anikonov, Yu. E. (1978). *Some Methods for the Study of Multidimensional Inverse Problems for Differential Equations.* Nauka. Novosibirsk (in Russian).

Atamanov, E. R. (1977). On uniqueness and stability of the multipoint problem solution for the pseudoparabolic equation. *Questions of Well-Posedness of Problems of Mathematical Physics.* Computer Center, Siberian Branch of USSR Academy Sci., Novosibirsk, 12-22 (in Russian).

Atamanov, E. R. (1979). Uniqueness and the stability estimate of one ill-posed problem solution for the pseudoparabolic equation. *Investigations on Integro-Differential Equations.* Frunze, **8**, 218-223 (in Russian).

Atamanov, E. R. (1983). The existence theorem for Cauchy problem solution for the pseudoparabolic equation with the data in the time axis. *Investigations on Integro-Differential Equations.* Frunze,**16,** 264-268 (in Russian).

Atamanov, E. R. and Mamayusupov, M. Sh. (1990). *Nonclassical Problems for Pseudoparabolic Equations.* Ilim, Frunze (in Russian).

Barenblatt, G. I. (1963). On some boundary-value problems for the liquid filtration equations in split media. *Appl. Mech. and Mat.* **27,** (2), 348-350 (in Russian).

Barenblatt, G. I., Zheltov, Yu. P., and Kochina P. N. (1960). On the base representations of filtration theory for homogeneous liquid in split media. *Appl. Mech. and Mat.* **24,** (5), 852-854 (in Russian).

Bukhgeim, A. Z. (1983). *Volterra Equations and the Inverse Problems.* Nauka, Novosibirsk (in Russian).

Calderon, A. (1958). Uniqueness in the Cauchy problem for partial differential equations. *Am. J. Math.* **80,** 16-36.

Carleman, T. (1933). Sur les systemes Lineaires aux derivées partielle du premier ordre a'deux variables. *Compt. Acad. Sci (Paris)* **197,** 471-474.

Carleman, T. (1939). Sur un problème d'unicité pour les systemes dequations aux derivées partielles a'deux varaibles. *Arkif. Math. Astr. Pys.* **26,** B:17.

Chen, P. and Gurtin, M. (1969). On a theory of heat conduction involving two temperatures. *Z. Angew. Math. and Phys.* **19,** 614-627.

Chen, P., Gurtin, M., and Wulliams, W. (1969). On the thermodynamics of nonsimple elastic material with two temperatures. *Z. Angew. Math. and Phys.* **20,** 107-112.

Colton, D. (1972a). On the analytic theory of pseudoparabolic equations. *Quart. J. Math.* **23,** 179-192.

Colton, D. (1972b). Pseudoparabolic equations in one space variable. J. *Different. Equat.* **12,** (3), 559-565.

Colton, D. and Wimp, J. (1979). Asymptotic bechaviour of the fundamental solution to the equation of the heat conduction in two temperatures. *J. Math. Anal. and Appl.* **69,** (2), 411-418.

Cordes, H. (1960). Uber die eindentige Bestimmtheit der Losungen elliptischer Differentialeleichum gen durch anfangsvorgaben. *Nachr. Akad. Wiss. Göttingen.* **11,** 239-256.

Douglis, A. (1953). Uniqueness in Cauchy problems for elliptic systems of equations. *Comm. Pure Appl. Math.* **6,** 291-298.

Dzertser, E. S. (1972). Generalization of the equation of ground water motion with free surface. *Dokl. Akad. Nauk SSSR* **232,** (5) (in Russian).

Evgrafov, M. A. (1968). *Analytic Functions.* Nauka, Moskow (in Russian).

Fedotov, A. M. (1982). *Linear Ill-Posed Problems with Random Errors in Data.* Nauka, Novosibirsk (in Russian).

Garipov, R. M. (1974). Non-hyperbolic boundary problem for the wave equation. *Dokl. Akad. Nauk SSSR* **219,** (4), 777-780 (in Russian).

Garipov, R. M. and Kardakov, V. V. (1973). The Cauchy problem for hyperbolic equation with nonspatial initial manifold. *Dokl. Akad. Nauk SSSR* **213,** (5), 10-12 (in Russian).

Gilbarg, D. and Trudinger, N. (1989). *Elliptic Differential Equations with Partial Derivatives of the Second Order.* Nauka, Moscow (in Russian).

Grasseli, M., Kabanikhin, S. I., and Lorenzi, A. (1992). The inverse problem for the integro-differential equation. *Sibirsk. Mat. Zh.* **33,** (3) (in Russian).

Gurtin, M. E. and Williams, M. (1967). An axiomatic foundation for continuim thermodynamics. *Arch. Ration. Mech.* **26,** 83-117.

Hadamard, J. (1932). *Le Problème de Cauchy et les Equations aux Derivées Partielles Lineaires Hyperboliques.* Hermann, Paris.

Hal'per, S. A. (1960). The Cauchy problem for general systems of linear equations with the partial derivatives. *Trudy Mosk. Mat. Obshch.* **9,** 401-423 (in Russian).

Hartman, P. and Witner, A. (1953). On the local behaviour of solutions of non-parabolic partial differential equations. *Amer. J. Math.* **75,** 449-476.

Holmgren. (1905). On Cauchy problem vid de lineara, purtiella different i alalekvationerna of 2: dra ordingen. *Arch. Math.* **2,** (4).

Hörmander, L. (1965). *Linear Differential Operators with Partial Derivatives.* Mir. Moscow (in Russian).

Imanaliev, M.I. (1977). *Methods for Solving Nonlinear Inverse Problems and their Applications.* Ilim. Frunze (in Russian).

Ivanov, V. K., Vasin, V. V., and Tanana, V. P. (1978). *Theory of Linear Ill-Posed Problems and their Applications.* Nauka, Moscow (in Russian).

John, F. (1960). Continuous dependence on data for solutions of partial differential equations with a prescribed bound. *Comm. Appl. Math.* (4), 551-585.

Kaistrenko, V. M. (1975). On the Cauchy problem of hyperbolic equation of the second order with data in the time-like surface. *Sibirsk. Mat. Zh.* **16,** (2), 395-398 (in Russian).

Landis, E. M. (1959). Some problems of the qualitative theory of elliptic and parabolic equations. *Uspekhi Mat. Nauk* **14,** (1), 21-86 (in Russian).

Landis, E. M. (1963). Some problems of the qualitative theory of elliptic equations. *Uspekhi Mat. Nauk* **18,** (1), 3-62 (in Russian).

Landis, E. M. (1972). On the behaviour of solution to the parabolic equation on the characteristic. *Mat. Zametki* **12,** (3), 257-262 (in Russian).

Lavrent'ev, M. M. (1956). On the Cauchy problem for the Laplace equation.*Izv. Akad. Nauk SSSR, Ser. Math.* **20,** 819-842 (in Russian).

Lavrent'ev M. M. (1962). *Some Ill-Posed Problems of Mathematical Physics.* Press of Siberian Branch of USSR Academy Sci., Novosibirsk (in Russian).

Lavrent'ev, M. M. (1977). *On Certain Problems of Ill-Posed Problem Theory of Mathematical Analysis.* Preprint. Computer Center, Siberian Branch of USSR Academy Sci., Novosibirsk (in Russian).

Lavrent'ev, M. M. and Amonov, B. K. (1976). Determination of the solution to the diffusion equation by the solution's values in the discrete sets. *Dokl. Akad. Nauk SSSR* **226,** (6), 1284-1285 (in Russian).

Lavrent'ev, M. M. and Amonov, B. K. (1982). Uniqueness and stability of an internal inverse problem for the heat conductivity equation. *Dokl. Akad. Nauk SSSR* **262,** (3), 528-530 (in Russian).

Lavrent'ev, M. A. and Lavrent'ev, M. M. (1976). On some inequalities for space harmonic functions. *Amer. Acad. Scient. Fenneix. Ser. A. I. Math.* (2), 303-306.

Lavrent'ev, M. M., Romanov, V. G., and Shishatskii, S. P. (1980). *Ill-Posed Problems of Mathematical Physics and Analysis.* Nauka, Moscow (in Russian). English transl. American Math Soc., Providence, Rhode Island (1986).

Levinson, N. (1949). The inverse Sturm-Liouville problem. *Math. Pidsskrift.* **13,** 25-30.

Lions, Zh. L. (1972). *Some Methods for Solving Nonlinear Boundary-Value Problems.* Mir, Moscow (in Russian).

Lorenzi, A. and Sinestrari, E. (1986). *An Inverse Problem in the Theory of Materials with Memory.* Dipartimento di matematica F. Enriques, Quaderno 23.

Markovitz, M. and Coleman, B. (1963). Incompressible second-order fluids. *Adv. Appl. Mech.* **8,** 69-101.

Markovitz, H. and Coleman, B. (1964). Nonsteady Helical Flows of Second-Order Fluids. *Phys. Fluids.* **7,** 833-841.

Matveev, N. M. and Sviriduk, G. A. (1985). *On a Linear Pseudoparabolic Equation for Nonstationary Filtration.*Mathematical Physics. Leningrad (in Russian).

Milne, E. H. (1926). The diffusion of trapped radiation through a gas. *J. London Math. Soc.* **1,** 40-51.

Mizohata, S. (1959). Le problème de Cauchy pour les systemes hyperboliques et paraboliques. *Met. Coll. Sci. Univ. Kyoto.* **32,** 181-212.

Mizohata, S. (1977). *The Theory of Equations with Partial Derivatives.* Mir. Moscow (in Russian).

Nirenberg, L. (1957). Uniqueness in Cauchy problem for differential equations with constant leading coefficients. *Comm. Pure Appl. Math.* **10,** 89-105.

Polubarinova-Kochina, P. Ya. (1977). *The Theory Of Ground Liquid Motion.* Nauka, Moscow (in Russian).

Protter, M. (1961). Properties of solutions of parabolic equations and inequalities. *Can. J. Math.* **13,** (2), 331-345.

Ptashnic, B. I. (1984). *Ill-Posed Boundary Problems for Differential Equations with Partial Derivatives.* Naukova Dumka, Kiev (in Russian).

Rao, R. C. and Ting, T. W. (1972). Solutions of pseudo-heat equations in the whole space. *Arch. Rat Mech. Anal.* **49,** 57-78.

Rao, R. C. and Ting, T. W. (1973). Initial-value problems for pseudoparabolic partial differential equations Indian. *Univ. Math. J.* **25.**

Romanov, V. G. (1973a). *Inverse Problems for Differential Equations.* Izdat. Novosibirsk. Gos. Univ., Novosibirsk (in Russian).

Romanov, V. G. (1973b). The abstract inverse problems and its correctness. *Funct. Anal. and its Appl.* **7,** (3), 67-74.

Romanov, V. G. (1984). *Inverse Problems of Mathematical Physics.* Nauka, Moscow (in Russian).

Romanov, V. G., Kabanikhin, S. I., and Pukhnacheva, T. P. (1984). *Inverse Problems of Electrodynamics.* Novosibirsk (in Russian).

Rubinstein, L. M. (1948). On a process of the heat propagation in heterogeneous media. *Izv. Akad. Nauk SSSR. Ser. Geograph. Geophys.* **12,** (1), 27-45 (in Russian).

Rundell, M. (1979). The uniqueness class for the Cauchy problem for pseudoparabolic equations. *Proc. Amer. Math. Soc.* **76,** (2), 254-257.

Rundell, M. (1980). Determination of an unknown nonhomogeneous term in a linear partial differential equation from overspecified boundary data. *Appl. Anal.* **10,** 231-242.

Rundell, M. and Stecher, M. (1977). Remarks concerning the supports of solutions of pseudoparabolic equations.*Proc. Amer. Math. Soc.* **63,** (1), 77-80.

Shirota, T. (1957). On Cauchy problem for linear partial differential equations with variable coefficients. *Osaka Math. J.* **97,** 43-60.

Shishatskii, S. P. (1973). *A priori* estimates in the problem of wave field continuation along cylindrical time-like surface. *Dokl. Akad. Nauk SSSR* **213,** (1), 49-50 (in Russian).

Shishatskii, S. P. (1975). On uniqueness of the Cauchy problem solution with the data in the time-axis interval for the degenerate parabolic equation. *Differential Equations.* **11,** (9), 1454-1459 (in Russian).

Shishatskii, S. P. (1977). On determining the function harmonic in the plane by its values in three parallel intervals. *Questions of Correctness of Mathematical Physics Problems.* Computer Center, Siberian Branch of USSR Academy Sci., Novosibirsk, 143-149 (in Russian).

Shishatskii, S. P. (1978). On uniqueness and stability of certain Cauchy problems for second-order equations. *Computing Methods in Mathematical Physics, Geophysics and Optimal Control.* Computer Center, Siberian Branch of USSR Academy Sci., Novosibirsk, 209-215 (in Russian).

Shishatskii, S. P. (1979). On zeros of solution to the mixed problem for a parabolic equation. *Ill-Posed Mathematical Problems and Problems of Geophysics.* Computer Center, Siberian Branch of USSR Academy Sci., Novosibirsk, 122-133 (in Russian).

Shishatskii, S. P. (1980). On the stability type of the Cauchy problems solution with data in the time half-axis for a pseudoparabolic equation. *Uniqueness, Stability and Methods for Solving Inverse and Ill-posed Problems.* Computer Center, Siberian Branch of USSR Academy Sci., Novosibirsk, 119-125 (in Russian).

Shishatskii, S. P. (1982). *The Uniqueness Problem for Cauchy Problem Solutions for Degenerate Equations.* Computer Center, Siberian Branch of USSR Academy Sci., Novosibirsk (in Russian).

Shishatskii, S. P. and Atamanov, E. R. (1978). *Uniqueness and the Stability Estimate of a Problem for a Pseudoparabolic Equation. Preprint.* Computer Center, Siberian Branch of USSR Academy Sci., Novosibirsk (in Russian).

Shkhanyukov, M. Kh. (1982). On some boundary-value problems for the third-order equation and its extremal properties. *Differential Equations.* **18,** (4) (in Russian).

Showalter, R. and Ting, T. (1970). Pseudoparabolic partial differential equations. *SIAM. J. Math. Anal.* **1,** 1-26.

Stoilov, S. (1962). *Complex Function Theory.* Izd. Inostr. Lit., Moscow, **2** (in Russian).

Suvejka, I. V. (1980). Mixed problems for one nonstationary equation. *Math. Research* **58,** Kishinev (in Russian).

Suvejka, I. V. (1983). On the Cauchy problem for nonstationary equations with the small viscosity. *Math. Anal.* (73), Kishinev, 89-100 (in Russian).

Sviriduk, G. A. (1985). *On Solvability and Regularity of Solutions to one Problem of Mathematical Filtration Theory.* Mathem. Phys., Leningrad (in Russian).

Tanana, V. P. (1981). *Methods for Solving Operator Equations.* Nauka, Moscow (in Russian).

Tikhonov, A. N. and Arsenin, V. Ya. (1974). *Methods for Solving Ill-Posed Problems.* Nauka, Moscow (in Russian).

Ting, T. W. (1963). Certain non-steady flows of second-order fluids. *Arch Rat. Mech. Anal.* **14,** 1-26.

Ting, T. W. (1969). Parabolic and pseudoparabolic partial differential equations. *J. Math. Soc. Japan.* **21**, 440-453.

Vodakhova, V. A. (1984). *Boundary-Value Problems for the Third-Order Equation of Mixed Pseudoparabolic Type.* Kandid. Dissert. Baku (in Russian).

Bei Fragen zur Produktsicherheit wenden Sie sich bitte an:
If you have any questions regarding product safety,
please contact:

Walter de Gruyter GmbH
Genthiner Straße 13
10785 Berlin
productsafety@degruyterbrill.com

संचयिता

रामधारी सिंह 'दिनकर'

भारतीय ज्ञानपीठ

पहला संस्करण : 1973
पाँचवाँ संस्करण : 1993
दसवाँ संस्करण : 2005
तेरहवाँ संस्करण : 2010
चौदहवाँ संस्करण : 2012
पन्द्रहवाँ संस्करण : 2015
सोलहवाँ संस्करण : 2017

लोकोदय ग्रन्थमाला : ग्रन्थांक 358

ISBN 978-93-263-5070-9

प्रकाशक :
भारतीय ज्ञानपीठ
18, इन्स्टीट्यूशनल एरिया, लोदी रोड, नयी दिल्ली-110 003

मुद्रक : चार दिशाएँ प्रिंटर्स, नोएडा

आवरण : अशोक भौमिक

आवरण साज-सज्जा : ज्ञानपीठ कला प्रभाग

SANCHAYIATA
(Hindi Poetry)
by Ramdhari Singh Dinkar

Published by
Bharatiya Jnanpith
18, Institutional Area, Lodi Road, New Delhi-110 003
Ph. : 011 24698417, 24626467; 23241619 (Daryaganj)
Mob. : 9350536020; e-mail : bjnanpith@gmail.com
sales@jnanpith.net, website : www.jnanpith.net

Sixteenth Edition : 2017
Price : Rs. 200

प्रस्तुति

(प्रथम संस्करण से)

भारतीय ज्ञानपीठ ने सदा प्रयत्न किया है कि पुरस्कार-समर्पण समारोह के अवसर पर पुरस्कृत कृति का हिन्दी अनुवाद प्रकाशित किया जाए, या यदि कृति हिन्दी की है तो पुरस्कृत साहित्यकार के रचना-संसार की एक झाँकी प्रस्तुत करने के लिए उसकी कृतियों से सृजेता द्वारा स्वयं चुनी गयी रचनाओं का संकलन प्रकाशित किया जाए।

बाँसुरी (शंकर कुरुप), गणदेवता (ताराशंकर वन्द्योपाध्याय), श्रीरामायण-दर्शनम् (अंशतः) (के.पी. पुट्टप्पा) एवं निशीथ (उमाशंकर जोशी), चिदम्बरा संचयन (विविध भारतीय भाषाओं में अनूदित—पन्त), बज़्मे ज़िन्दगी रंगे शाइरी (फ़िराक़), स्मृति सत्ता भविष्यत् तथा अन्य कविताएँ (विष्णु दे)—ये सब कृतियाँ समारोह-आयोजन के अंग बनकर प्रकाशित हुई हैं। निःसन्देह हिन्दी साहित्य इन प्रकाशनों से समृद्ध हुआ है और भारतीय ज्ञानपीठ ने इनके माध्यम से एक महत्त्वपूर्ण आनुषंगिक उपलब्धि पायी है।

1 दिसम्बर 1973 को आयोजित पुरस्कार समारोह के लिए जब पुरस्कृत कृति 'उर्वशी' के रचयिता दिनकरजी के कृतित्व के प्रकाशन का प्रश्न आया तो हमने दिनकरजी से अनुरोध किया कि इस अवसर के लिए वह एक ऐसा काव्य-संकलन तैयार कर दें जो उनके 'सर्वश्रेष्ठ' का प्रतिनिधित्व करता हो, और इस रूप में जिसका चुनाव मात्र उन्हीं के द्वारा हुआ हो। दिनकरजी ने हमारे अनुरोध को मान दिया, और अब यह 'संचयिता' इस रूप में आपके हाथों में है। इस संकलन का प्रकाशनोत्सव भी पुरस्कार-समारोह के मंच से हुआ है, अतः यह कृति अपने आप में दिनकरजी के जीवन और कृतित्व के इतिहास की अविस्मरणीय कड़ी बन गयी है।

अब तक प्रकाशित लगभग तीस काव्य-कृतियों, संकलनों आदि में से श्रेष्ठतम चुनना बहुत कठिन है, किसी के लिए भी, स्वयं कवि के लिए तो असम्भव ही है क्योंकि प्रत्येक रचना उसकी आत्मजा है; वह किसे कहे कि यह वरणीय नहीं। दिनकरजी के लिए समस्या कितनी भी कठिन क्यों न रही हो, हमें इसे 'श्रेष्ठतम' कहने में संकोच इसलिए नहीं कि चक्रवाल (1956), कविश्री (1957) और लोकप्रिय कवि दिनकर (1960) जैसे संकलन पहले प्रकाशित हो चुके हैं, और दिनकरजी ने अपनी कविता के मान-प्रतिमान निश्चित करने में प्रत्यक्ष-अप्रत्यक्ष रूप से प्रकाशकीय सहयोग दिया होगा। 'संचयिता' कृति की विशेषता यह है कि यह न केवल पिछले एक दशक की कविताओं की श्रेष्ठता को समाहित कर रही है, प्रत्युत 'श्रेष्ठों' के संकलनों में से 'श्रेष्ठतर' का यह संकलन है।

दिनकर का काव्य विस्तार की दृष्टि से, कालावधि को परिमापित करने की दृष्टि से, भाषागत, विषयगत प्रयोगों की दृष्टि से, गुणगत और प्रकृतिगत विविधता की दृष्टि से, तथा सर्वोपरि, साहित्यिक प्रभाव की दृष्टि से अद्वितीय है, अद्‌भुत है, महत् है। पुरस्कार समारोह में समर्पित प्रशस्ति का साक्ष्य इस प्रकार है :

''श्री दिनकर ने छायावाद की अस्पष्ट और वायवीय विषय-वस्तु तथा रूप विधानोंवाली काव्य-परम्परा से अलग होकर आधुनिक हिन्दी कविता को एक ऐसी ओजमयी ऋजु भाषा-शैली दी जो पुनरुत्थानशील राष्ट्रीयता की अदम्य प्रेरणाओं को अभिव्यक्त करने में समर्थ हुई। वे हिन्दी साहित्य में एक अपूर्व घटना-तथ्य जैसे बनकर आये, क्योंकि जिस सशक्तता से उनकी लेखनी ललकार और विद्रोह का झण्डा ऊँचा कर सकी, उसी से सुशान्त चिन्तन और गीतात्मक भाषा-शैली में मानव-मन के कोमल भावों को प्रकट करती आयी।

उनके काव्य में इन दोनों प्रवृत्तियों का बार-बार बड़ा मुग्धकर पुनरावर्तन होता है। कोई आश्चर्य नहीं कि अचूक आह्वान-शक्ति और भावात्मक प्रकृति दोनों के स्वभाव के कारण उनके काव्य को 'दहकते अंगारों पर इन्द्रधनुषों की क्रीड़ा' से उपमित किया गया है।

पुरस्कार-जयी काव्य-नाटक 'उर्वशी' फल-सिद्धि है कवि की प्रतिभा और एक चिन्तक के प्रेमभाव के अर्थ, उद्देश्य और विस्तारों की अन्वेषणा की; मानव की ऐन्द्रिय जगत् से होते चरम परम की शोध की। तीस काव्यकृतियों और साहित्य अकादमी द्वारा पुरस्कृत 'संस्कृति के चार अध्याय' सहित, निबन्ध,

कहानी, संस्मरण-दैनिकी, समीक्षा-इतिहास आदि विभिन्न साहित्य-विधानों की पचीस प्रांजल गद्य कृतियों के प्रातिभ प्रणेता दिनकरजी इसी प्रकार दर्शन और विश्वकाव्य की गहराइयों में पैठते हुए साहित्यिक उपलब्धि के उत्तरोत्तर भव्य शिखरों तक पहुँचने की अविराम आशाएँ बँधाये हुए हैं।''

दिनकर की कविता का विकास आदि से अन्त तक द्वन्द्वों से जूझने की कथा है—द्वन्द्व जो प्रबुद्ध-चेता कवि को चिन्तन और भावना के क्षेत्रों में, प्रत्येक स्तर पर घेरते हैं, भरमाते हैं, विकल करते हैं और चुनौती देकर उसे साधना एवं समाधान के प्रगति-पथ पर ले चलते हैं। छायावाद की आभिजात्य शैली और रहस्यात्मक कथ्य से द्वन्द्व 'रेणुका' (1935) के सहज रचना-शिल्प और जन-सुबोध, सामयिक कथ्य का; 'हुंकार' (1938) के क्रान्तिकारी आह्वान से द्वन्द्व 'कुरुक्षेत्र' (1946) का; और स्वयं 'कुरुक्षेत्र' है एक सर्वाधिक समर्थ अभिव्यक्ति उस द्वन्द्व की जो युद्ध और शान्ति, हिंसा और अहिंसा, प्रवृत्ति और निवृत्ति की जीवन-पद्धति तथा विज्ञान और आत्मज्ञान की परिणति में निहित है। बीच में 'द्वन्द्वगीत' (1940) तो अभिधा में ही स्पष्ट है। 'रश्मिरथी' (1952) के प्रतिपक्ष में आ खड़ी होती है *'उर्वशी'* (1961)—'उर्वशी' जो अप्सरा और लक्ष्मी, द्विधाग्रस्त मानव और द्विधामुक्त देवता, एवं काम और अध्यात्म के द्वन्द्वों की गाथा है। रहट के घटों की माला के उतार-चढ़ाव की भाँति 'चक्रनेमि-क्रमेण' 'उर्वशी' की प्रतिगामी दिशा की प्रभावकारी मुद्रा 'परशुराम की प्रतीक्षा' (1963) में परिलक्षित है।

कड़ी-कड़ी मिलकर जब द्वन्द्वों की शृंखला एक तरह से पूरी हो गयी तो जैसे आगे चरण-निक्षेप कठिन हो गए। या, आगे की यात्रा के लिए शक्ति-संचय का दशक जैसे जरूरी हो गया। अर्थात् रश्मिरथी से उर्वशी तक आने में एक दशक लगा। बीच के व्यवधान को विश्रान्ति की कविताओं ने पाटा—'दिल्ली' और 'नीम के पत्ते' (1954), 'सूरज का ब्याह' और 'नीलकुसुम' (1955), और 'नये सुभाषित' (1957) ने 'परशुराम की प्रतीक्षा' के बाद 'दिनकर के गीत' (1973) तक का एक दशक और इसी प्रकार विश्रान्ति की कविताओं का काल है—'कोयला और कविता', 'मृत्ति तिलक', 'दिनकर की सूक्तियाँ' और 'आत्मा की आँखें' (सब 1964 में प्रकाशित)।

इन दो दशकों का तथाकथित, तथा-परिलक्षित, 'विश्रान्ति-काल', दिनकर के लिए घोर आन्तरिक संघर्षों का काल रहा है—संघर्ष, जिसने दिनकर के बाहर के प्रसार को अन्तर्मुखी किया, जिसने उन्हें नयी उपलब्धियों से मण्डित किया—विश्व कविता का अध्ययन और प्रभाव, 'नीलकुसुम' (1955), 'सीपी और शंख' (1957),

'काव्य की भूमिका' (1958), 'शुद्ध कविता की खोज' (1966); दार्शनिक ग्रन्थों का मनन, इतिहास का अध्ययन, राष्ट्र की आत्मा को परिभाषित करने का प्रयत्न 'संस्कृति के चार अध्याय' (1956), 'धर्म, नैतिकता और विज्ञान' (1959), इन अन्तरालों में दिनकर के चिन्तन की गरिमा गद्य के निखार में व्यक्त हुई।

इन-दो दशकों की विराम-कथा में जो बात छूट गयी, वही शायद अब सबसे महत्त्वपूर्ण है—'हारे को हरिनाम' (1970)। क्या आन्तरिक संघर्ष ने दिनकर को तोड़ दिया ? या, नीचे की कर्दम और ऊपर की धूल से ऊँचे उठकर, त्वचा से छिटककर, जीने की साधना का यह प्रवेश-द्वार है ? शायद दोनों ही। क्योंकि द्वन्द्व दिनकर के जीवन के श्वास-प्रश्वास हैं।

दिनकर के काव्य के उतार-चढ़ाव के प्रत्येक चरण को उनके जीवन के विकास-क्रम, या कहें घटना-क्रम के साथ जोड़ा जा सकता है, जोड़ा जाना चाहिए। जन्म-भूमि सिमरिया (मुँगेर, बिहार) के बाँस-वन, प्राकृतिक सुषमा, और साथ-साथ गाँववालों के जीवन का विकराल दैन्य, अशिक्षा और मरण की देहरी पर नवजन्म के गीत। इतिहास में आनर्स करने के बाद का अकिंचन अध्यापक, फिर एक राष्ट्रधर्मी, स्वाभिमानी सब रजिस्ट्रार जिसकी सेवा के प्रथम चार वर्षों में ब्रिटिश शासन ने उसे 22 बार ट्रांसफर किया कि 'हिमालय' और 'हुंकार' का यह कवि नौकरी छोड़ दे। देश की स्वतन्त्रता के उपरान्त पटना में छह वर्ष तक सरकारी प्रचार-विभाग का अधिकारी; फिर कॉलेज में हिन्दी विभाग का अध्यक्ष-प्राध्यापक; विश्वविद्यालय का कुलपति। बीच-बीच में अपने पुत्रों के प्रकाशनगृह के लिए विश्रान्ति काल की इन कविताओं के योगदान की सम्भावना बनाते गये। दिनकर ने अपने कार्यकाल के 38 वर्षों में लगभग 19 वर्ष गाँवों में और 19 वर्ष दिल्ली में बिताये हैं। संसद् सदस्य के रूप में 12 वर्ष, और हिन्दी सलाहकार अधिकारी के रूप में 7 वर्ष। उनके अनुभवों की व्यापकता और गहनता के क्या कहने ! संसद् सदस्य की हैसियत से दिल्ली में रहने के गौरव ने नये द्वन्द्वों को जन्म दिया। यहाँ रहकर भारत-भाग्य विधाताओं के खोखले जीवन की अन्तरंग झाँकी; शासकीय दाँव-पेंच, उखाड़-पछाड़, जनता के प्रति हृदयहीन उपेक्षाभाव, क्रोड़ों के व्यय के आँकड़ों में जीवन की कंगाली के छाया-चित्रों का उपहास और त्रास दिनकर ने अभिव्यक्त किया :

"मैं भारत के रेशमी नगर में रहता हूँ।
जनता तो चट्टानों का बोझ सहा करती
मैं चाँदनियों का बोझ किसी विध सहता हूँ।

गन्दगी, गरीबी, मैलेपन को दूर रखो
शुद्धोदन के पहरेवाले चिल्लाते हैं,
है कपिलवस्तु पर फूलों का शृंगार पड़ा
रथ-समारूढ़ सिद्धार्थ घूमने जाते हैं।

रेशमी कलम से भाग्य-लेख लिखने वालो
तुम भी अभाव से कभी ग्रस्त हो रोये हो ?
बीमार किसी बच्चे की दवा जुटाने में
तुम भी क्या घर पर पेट बाँधकर सोये हो ?''

दिनकर धन्य हैं कि उन्हें जीवन में वह सब मिला जिसकी भाग्यवान् लोग कामना किया करते हैं। यशस्वी कवि जिसे जनता ने प्यार किया, सचमुच उसकी कविता से साक्षात्कार करके उससे प्रभावित होकर। पुरस्कार और पदक और पद और पद्मभूषण। और, प्रेम...कौन कवि लिख सकता था 'उर्वशी' बिना प्रेम के अगोचर छोरों को छुए ?

''मैं तुम्हारे बाण का बींधा हुआ खग,
वक्ष पर धर शीश मरना चाहता हूँ।
मैं तुम्हारे हाथ का लीला कमल हूँ
प्राण के सर में उतरना चाहता हूँ।

मैं तुम्हारे रक्त के कण में समाकर
प्रार्थना के गीत गाना चाहता हूँ।''

सौभाग्य की मंजूषा रीती न रह जाए इसलिए दिनकर को विधाता ने वेदनाओं से मथा है : तरुण पुत्र की मृत्यु, गृह-उच्छेद, विषण्णता, राग-उद्वेग और घोर चिन्ता !

''हृदय छोटा हो,
तो शोक वहाँ नहीं समाएगा।
और दर्द दस्तक दिये बिना लौट जाएगा।

टीस उसे उठती है
जिसका भाग्य खुलता है

वेदना गोद में उठाकर
सबको निहाल नहीं करती,
जिसका पुण्य प्रबल होता है
वही अपने आँसुओं से धुलता है।''

कौन जानता है कामना-पूर्ति के इति पर चरण रखकर दिनकर आगे बढ़ जाएँ और 'सिमरिया के सन्त' से एकाकार हो जाएँ।

दिल्ली
20 नवम्बर, 1973

–लक्ष्मीचन्द्र जैन

आत्मकथन

(पुरस्कार-समर्पण के अवसर पर)

ज्ञानपीठ पुरस्कार ने एक शिखर तैयार कर दिया है, जिस पर खड़ा होने से आदमी सारे देश को दिखाई पड़ जाता है और उत्सव के दिन क्षण भर को उसे यह अधिकार भी मिल जाता है कि वह कुछ बोले और लोग उसकी बातों को सुनें। यह बहुत बड़ा गौरव है और मैं सच्चे हृदय से उन सभी सहृदय विद्वानों को धन्यवाद देता हूँ, जिन्होंने मेरे 'उर्वशी' काव्य को इस पुरस्कार के योग्य समझा और मुझे इस उच्च शिखर पर खड़ा होने का अवसर प्रदान किया है। जिस विशाल देश में सोलह भाषाएँ संविधान द्वारा स्वीकृत हों और प्रायः सभी भाषाओं में साहित्य-निर्माण का कार्य हो रहा हो, उस देश में साहित्य के क्षेत्र में सार्वदेशिक मंच का निर्माण बड़ा ही कठिन कार्य है। जब भारतीय ज्ञानपीठ ने अपने पुरस्कार की योजना पहले-पहल विज्ञापित की थी, उस समय पुरस्कार-योजना के शुभचिन्तक भी मन ही मन सहमे हुए थे। उन्हें चिन्ता थी कि सभी भाषाओं की पुस्तकों के बीच तुलना का कार्य किस प्रक्रिया से सम्पन्न किया जाएगा और तुलना करने के पश्चात् जब प्रवर-समिति किसी ग्रन्थ को सर्वश्रेष्ठ घोषित करेगी, तब प्रवर-समिति का निर्णय सामान्यतः मान्य समझा जाएगा या नहीं। किन्तु पिछले आठ वर्षों के अनुभव ने यह सिद्ध कर दिया है कि यदि सदाशयता हो और पूरा परिश्रम किया जाए, तो यह असम्भव कार्य भी सम्भव बनाया जा सकता है।

भारतीय ज्ञानपीठ ने जिस बीहड़ काम में हाथ डाला था, उसमें उसे सफलता प्राप्त हुई। इसलिए वह देश में प्रशंसा का पात्र हो गया और पिछले आठ वर्षों में इस पर प्रशंसा के जो पुष्प बरसे हैं, ज्ञानपीठ की सेवा उनके योग्य थी। मेरी दृष्टि में ज्ञानपीठ की इस सफलता का ऐतिहासिक महत्त्व है। स्वराज्य होने के पूर्व हम अपने देश की अन्य भाषाओं के साहित्यकारों के नाम तब तक

नहीं सुनते थे, जब तक वे रवीन्द्र, प्रेमचन्द, शरत् या इक़बाल न हो जाएँ। स्वराज्य होने के बाद भारतीय भाषाओं को परस्पर समीप लाने का आन्दोलन आरम्भ हुआ और हमारी प्रत्येक भाषा के लोग अपने देश की अन्य भाषाओं के श्रेष्ठ लेखकों के नाम कुछ ज्यादा सुनने लगे। इस दिशा में साहित्य अकादमी ने जो कार्य किया है, उसकी महिमा स्पष्ट है। किन्तु इतना कुछ होने पर भी साहित्य में अखिल भारतीय मंच की कल्पना निराकार की निराकार ही रही। उसे साकार करने का श्रेय इतिहास भारतीय ज्ञानपीठ को देगा, इसमें मुझे तनिक भी सन्देह नहीं है। भारतीय ज्ञानपीठ ने भारत-राष्ट्र का एक बहुत बड़ा कार्य कर दिया, जिसके लिए मैं उसे बधाई देता हूँ।

ज्ञानपीठ-पुरस्कार से केवल साहित्य की ही सेवा नहीं हो रही है, उससे भारत की भावात्मक एकता में भी वृद्धि हो रही है। भारत अपने बौद्धिक व्यक्तित्व को भी ऊपर उठा रहा है, साहित्य के क्षेत्र में उसके आत्मविश्वास में भी वृद्धि हो रही है और, अप्रत्यक्ष रूप से, वह अँग्रेजी के दबाव से भी निकलता जा रहा है।

ज्ञान का साहित्य मनुष्य किसी भी भाषा में लिख सकता है, जिससे उसने भलीभाँति सीख लिया हो, किन्तु रस का साहित्य वह केवल अपनी भाषा में रच सकता है। प्रत्येक ऐसे देश की आत्मा, जिसका कोई इतिहास है, उस देश की अपनी भाषा में बोलती है। भारत की भाषा उसकी अपनी भाषा है। अँग्रेजी के माध्यम से हमने भारत की महिमा का प्रचार किया है। अन्यथा अँग्रेज और अँग्रेजी के युग में भी भारत की आत्मा की भाषा उसकी अपनी ही भाषा थी। भारतीय भाषाओं का पुजारी होने के कारण भारतीय ज्ञानपीठ श्रद्धा का पात्र है। मैं श्रद्धा के भाव से उसे प्रणाम करता हूँ। अपनी भाषा की उन्नति करना और उन्नत देशों की भाषाएँ सीखना, ये दोनों परस्पर विरोधी कार्य नहीं हैं। किन्तु भारत को यदि मौलिक राष्ट्र बनाना है, तो उसकी अपनी भाषाओं को सर्वाधिक महत्त्व देना ही पड़ेगा।

अब मैं असमंजस में हूँ कि ज्ञानपीठ को धन्यवाद देने के पश्चात् आज मुझे बोलना क्या चाहिए। देश की रक्षा के लिए किया जाने वाला संघर्ष बड़ा, परिवार की रक्षा के लिए किया जानेवाला संघर्ष छोटा होता है। फिर भी मैंने सारी आयु इसी छोटे संघर्ष में बितायी है। दिन का जो ताजा हिस्सा था, वह परिवार के लिए रोटी कमाने में गया। उसके बाद जो अवसाद की घड़ियाँ होती हैं, उन्हीं में मैंने साहित्य की साधना की है। फिर भी साहित्य-संसार ने मेरी ओर

आँख उठाकर देखने की कृपा की, इसे मैं अपना सौभाग्य और उस प्रभु का वरदान समझता हूँ जिसकी कृपा से मूक बोलते और पंगु पर्वतारोहण करते हैं।

लगता है, पृथ्वी पर आने के पूर्व जब भगवान को प्रणाम करने गया, वे कलाकारों के बीच छेनी, टाँकी, हथौड़ी, कूँची और रंग बाँट रहे थे। लेकिन भगवान ने मुझे छेनी, टाँकी और हथौड़ी नहीं दी, जो पच्चीकारी के औजार हैं। उनके भण्डार में एक हथौड़ा पड़ा हुआ था। भगवान ने वही हथौड़ा उठाकर मुझे दे दिया और (जरा-सी आत्मश्लाघा के लिए क्षमा कीजिए) कहा कि जा, तू इस हथौड़े से चट्टान का पत्थर तोड़ेगा और तेरे तोड़े हुए अनगढ़ पत्थर भी काल के समुद्र में फूल के समान तैरेंगे।

लगता है, जब मैं हथौड़ा लेकर चला, मैं छेनी और टाँकी की ओर मुड़-मुड़कर लोभ से देख रहा था। वह लोभ मुझे जीवन-भर सताता रहा है और जीवन-भर मैं इस विचिकित्सा में पड़ा रहा हूँ कि कविता का वास्तविक प्रयोजन क्या है। क्या वह मनुष्य को जगाने, सुधारने और उन्नत बनाने के लिए है या उसका काम आदमी को रिझाना और उसे प्रसन्न करना है। या इनमें से कोई भी ध्येय कविता का ध्येय नहीं है ? जैसा कि एज़रा पॉउण्ड ने कहा है, कविता केवल कविता है, जैसे वृक्ष केवल वृक्ष है। वृक्ष अपनी जगह पर स्थिर खड़ा है। वह किसी को भी नहीं बुलाता। फिर भी लोग उसकी हरियाली को देखकर खुश होते हैं, उसकी छाया में बैठते हैं और पेड़ अगर फलदार हुआ, तो वे फलों को तोड़कर खा लेते हैं। चेतना के तल में जो घटना घटती है, जो हलचल मचती है, उसे शब्दों में अभिव्यक्ति देकर हम सन्तोष पाते हैं। यही हमारी उपलब्धि है। यदि देश और समाज को उससे कोई शक्ति प्राप्त होती है, तो वह अतिरिक्त लाभ है। किन्तु, इतना जरूर है कि पेड़ मनुष्य और पक्षियों के नहीं रहने पर भी फूल और फल सकते हैं, किन्तु पाठक और श्रोता न रहें, तो कवि कविता लिखेगा या नहीं, इसमें मुझे भारी सन्देह है। अगर सारी दुनिया खत्म हो जाए और केवल एक आदमी जीवित खड़ा हो, तो कवि होने पर भी, वह कविता शायद ही लिखेगा। मेरी दृढ़ धारणा है कि कविता व्यक्ति द्वारा सम्पादित सामाजिक कार्य है और शुद्ध कविता भी समाज के लिए ही लिखी जाती है।

अपने निर्माण के दिनों में प्रत्येक नये कवि को उस एक महाकवि का पता लगाना पड़ता है, जिसके समान वह बनना चाहता है। मेरा दुर्भाग्य या सौभाग्य यह रहा है कि मैंने एक के बदले ऐसे दो महाकवियों का पता लगा लिया, जिनके समान बनने की मुझमें उमंग थी। इनमें से एक थे श्री रवीन्द्रनाथ ठाकुर, जिनके

नाम की सारे संसार में धूम थी और जिनके प्रभाव में आकर भारत की कई भाषाओं में छोटे-छोटे रवीन्द्रनाथ पैदा हो गये थे; और दूसरे थे सर मोहम्मद इक़बाल, जिन्हें नोबेल पुरस्कार तो नहीं मिला था मगर जिनकी कविताएँ पाठकों के रुधिर में आग की तरंगें उठाती थीं, मन के भीतर चिन्तन का द्वार खोल देती थीं।

प्रभाव तो इन दोनों कवियों का मुझ पर पहले ही पड़ गया। यह पता बहुत बाद को चला कि रवीन्द्र और इक़बाल दो ध्रुवों के कवि हैं और वे अकसर आमने-सामने के दो विरोधी क्षितिजों से बोलते हैं। भगवान के प्रति रवीन्द्रनाथ का भाव सम्पूर्ण आत्म-समर्पण का भाव था–

प्रभु, तोमा लागि आँखि जागे।
देखा नाइ पाइ, शुधू पथ चाइ,
सेओ मोने भालो लागे।

किन्तु इक़बाल भगवान के उस प्रकार के भक्त थे जो चाहता है कि मैं भगवान में लीन नहीं होऊँगा, भगवान को ही मुझमें विलीन होना पड़ेगा–

खुदी को कर बुलन्द इतना
कि हर तक़दीर से पहले
खुदा बन्दे से खुद पूछे,
बता, तेरी रज़ा क्या है।

यह भी कि रवीन्द्रनाथ उपयोगिता को कोई महत्त्व नहीं देते थे। जहाँ तक उपयोगिता का सवाल है, आदमी और जानवर में कोई भेद नहीं है। मनुष्य का असली व्यक्तित्व तब बनता है, जब वह उपयोगिता के घेरे को लाँघकर ऐसी भूमि में पहुँच जाता है जो निरुद्देश्य आनन्द की भूमि है, जहाँ मनुष्य आत्मरक्षा की भावना से प्रेरित होकर साहित्य का सृजन नहीं करता, न समाज के किसी स्थूल प्रयोजन की पूर्ति के लिए कविता या चित्र बनाता है।

रवीन्द्रनाथ की तरह इक़बाल भी मानते हैं कि कला व्यक्तित्व की अभिव्यक्ति है। किन्तु, व्यक्तित्व की परिभाषा उनकी कुछ और है। जो अभी आराम में है, जो संघर्ष से दूर है, जो बड़े मक़सदों को हासिल करने के लिए जद्दोजहद नहीं कर रहा है, इक़बाल उस मनुष्य को व्यक्तित्वहीन समझते हैं। व्यक्तित्व की स्थिति संघर्ष की स्थिति होती है, तनाव की स्थिति होती है, और जो आदमी जितने ही अधिक तनाव में है, उसका व्यक्तित्व भी उतना ही बड़ा और बलवान है।

शुरू में ही मुझ पर रवीन्द्र और इक़बाल का जो विरोधी प्रभाव पड़ गया, उसके कारण मैं काफी वर्षों तक बेचैन रहा। मेरे निर्माण का समय वह था जब गाँधी जी समस्त देश को जीवन, जागरण, प्रेरणा और संघर्ष से आलोड़ित कर रहे थे। ऐसा समय क्या कोमल, वायवीय, निरुद्देश्य गीतों का समय होता है ? अथवा सम्भव है कि पराधीनता का विरोध, शोषण और साम्राज्यवाद पर प्रहार तथा समता के समर्थन में गान मैंने भी प्रचार के लिए नहीं, बल्कि इसलिए किया था कि वैसा करना मुझे अच्छा लगता था, आनन्ददायी मालूम होता था। कॉलेज में वर्ड्स्वर्थ की ये पंक्तियाँ कदाचित मैंने भी पढ़ी होंगी :

The gods approve the depth
And not the tumult of the soul

लेकिन जवानी भर मुझे इसकी परवाह ही नहीं रही कि देवताओं की पसन्द क्या है। मुख्य बात यह थी कि गर्म लोहे पर हथौड़े की चोट जोर से पड़ती है या नहीं। मैं रन्दा लेकर काठ को चिकनाने नहीं आया था। मेरे हाथ में तो कुल्हाड़ी थी। मैं जड़ता की लकड़ियों को फाड़ रहा था।

लेकिन मुझे राष्ट्रीयता, क्रान्ति और गर्जन-तर्जन की कविताएँ लिखते देखकर मेरे भीतर बैठे हुए रवीन्द्रनाथ दुःखी होते थे और संकेतों में कहते थे, "तू जिस भूमि पर काम कर रहा है, वह काव्य के असली स्रोतों के ठीक समीप नहीं है।" तब मैं 'असमय आह्वान' में 'हाहाकार' में तब अन्य कई कविताओं में अपनी किस्मत पर रोता था कि हाय, काल ने इतना कसकर मुझे ही क्यों पकड़ लिया ? मेरे भीतर जो कोमल स्वप्न हैं, वे क्या भीतर मुरझाकर मर जाएँगे ? उन्हें क्या शब्द बिलकुल ही नहीं मिलेंगे ?

लेकिन शब्द इन कोमल स्वप्नों को भी मिले। 'रसवन्ती' और 'द्वन्द्व-गीत' इन्हीं कोमल कविताओं के संग्रह हैं। किन्तु विरुद मेरा चारण और वैतालिक का ही रहा। 'हुंकार' के आमुख में मैंने स्वयं स्वीकार किया था :

अमृत-गीत तुम रचो कलानिधि,
बुनो कल्पना की जाली।
तिमिर-ज्योति की समर भूमि का
मैं चारण, मैं वैताली।

तब सन् 1943 के आसपास मेरा परिचय 'अदृश्य कवि' इलियट की कविताओं से हुआ। यह मेरी काव्य-चेतना में आनेवाला पहला भूडोल था। पूर्व इसके कि युग की बीमारी अपनी हस्ती का ऐलान करे, कवि को चाहिए कि वह

युग से कह दे कि तुम बीमार हो या होने वाले हो। इलियट की कविताएँ पूरी तरह मेरी समझ में नहीं आयीं, लेकिन तब भी मैं मान गया कि उन्होंने युग को यह चेतावनी दे दी है।

मैं बड़ी ही निश्चिन्तता और आत्मविश्वास के साथ गाता जा रहा था, साम्राज्यवाद की कुरूप छाती में अपने गीतों के खंजर चुभोता आ रहा था, पराधीनता की बेड़ियों पर उस हथौड़े से प्रहार करता आ रहा था, जो मुझे भगवान से प्राप्त हुआ था। किन्तु, इलियट को पढ़ते ही मैं थोड़ी देर के लिए ठिठककर रह गया। अरे, इलियट की कविताएँ हम लोगों की कविताओं यानी मेरे गुरु रवीन्द्रनाथ और इक़बाल की कविताओं से कितनी भिन्न हैं ! फिर मन ने कहा यह अवश्य ही परिस्थितियों का भेद है। इलियट उस दुनिया के कवि हैं, जो दुनिया समृद्धि की अधिकता से बेजार है, जिस दुनिया ने आत्मा को सुलाकर शरीर को जगा लिया है। किन्तु, हम तो पराधीन देश के कवि हैं। हमारा तो कोई देश ही नहीं है फिर हम 'मरु देश' की कल्पना कैसे कर सकते हैं ?

मगर इलियट को मैं चाहे जितना भी भूलना चाहता, मैं उन्हें भूल नहीं पाता था। उनकी कविताएँ समझ में भले ही नहीं आती हों, किन्तु वे मेरी शान्तिभंग करने में समर्थ थीं, मेरे मन को वे अकसर उस दिशा में भेज देती थीं जिस दिशा में कहीं कोई क्षितिज नहीं था, न कोई किताब खुलकर बन्द होती थी। मेरी चेतना के घाट बँध चुके थे, मेरी चमड़ी मोटी हो चुकी थी, मेरे मुहावरे अब बदले नहीं जा सकते थे। अतएव, इलियट के लिए यह असम्भव कार्य था कि वे मुझे बदलकर अपनी राह पर लगा लें। लेकिन मन बराबर यह महसूस करता रहा कि इलियट रवीन्द्र और इक़बाल से छोटे हों या बड़े, यह अलग बात है, किन्तु वे उन दोनों से भिन्न हैं और उनके साथ कविता में कोई ऐसी अदा उतरी है, जो संसार में और कभी दिखाई नहीं पड़ी थी। उस समय मैं यह क्या जानता था कि नवीनता का पाठ कविता ने इलियट में ही आकर नहीं पढ़ा ! यह पाठ इलियट की कविताओं से कोई पचास वर्ष पूर्व वह फ्रान्स में पढ़ चुकी थी। लेकिन भारतवासियों को तो भारत से बाहर केवल अँग्रेज दिखाई पड़ते थे। अतएव जब तक अँग्रेज नहीं बदले, भारतवासियों ने कविता के क्षेत्र में बदलने की बात सोची भी नहीं, जो बिल्कुल स्वाभाविक बात थी।

आगे चलकर श्री अरविन्द के एक लेख में मैंने पढ़ा कि भविष्य की कविता मन्त्र के समान छोटी और वैसी ही प्रभावशालिनी होगी। इलियट की कविता मन्त्र कविता का पूर्वाभास है। कविता जिस साधना में लगी हुई है, उसमें यदि वह

सफल हो गयी, तो मन्त्र की तरह संक्षिप्त होना उसका स्वभाव हो जाएगा और वह संकेत में इस प्रकार बोलेगी मानो आगामी पीढ़ियों को कवि चिट्ठी नहीं, तार भेज रहा हो। मुझे उन कवियों और आलोचकों पर तरस आता है जो इलियट की गणना साहित्य के लकड़बग्घों में करते हैं।

रवीन्द्र और इक़बाल ने मेरे हृदय-सरोवर को खूब हिलकोरा था। जब सरोवर किंचित् जड़ होने लगा, उसे इलियट और उनके समानधर्मी कवियों ने फिर से हिलकोर दिया। नयी कविता से मेरे घबराने का एक कारण यह था कि वह मेरी समझ में नहीं आती थी। दूसरे, उसने छन्द की राह छोड़ दी थी। किन्तु जब मैंने देखा कि चित्रकारी बालू और कोलतार से तथा मूर्तिकारी लोहे के तारों से की जा रही है, तब मैंने भी यह मान लिया कि कविता का गद्य में लिखा जाना कोई अनुचित बात नहीं है।

मुझे जो कुछ बनना था, रवीन्द्र और इक़बाल की कृपा से मैं बन चुका था। अब मुझे यह उम्मीद नहीं थी कि मुझ पर कविता के नये आन्दोलन का भी प्रभाव पड़ेगा। लेकिन जो बात जवानी में नहीं हो पायी, वह बुढ़ापे में आकर हो गयी। इस प्रभाव का अदृश्य आरम्भ 'नील कुसुम' में हुआ, उसका दृश्य प्रमाण 'हारे को हरि नाम' में मौजूद है। धर्म में निराकार से साकार की ओर गया हूँ। कविता में मेरी यात्रा साकार से निराकार की ओर है। पहले मैं यह जानता था कि कविता कहाँ से आ रही है और वह किस तरफ को जाएगी। अब मुझे यह मालूम ही नहीं होता कि कविता कहाँ से आती है और क्या उसका गन्तव्य है।

कविता युग के संचित ज्ञान का आख्यान नहीं है। कविता का क्षेत्र ज्यों-ज्यों नवीन होता जाता है, कवि त्यों-त्यों अधिक गहराई में उतरता जाता है, और ज्यों-ज्यों वह गहराई में उतरता जाता है, त्यों-त्यों यह बताने में वह अधिक असमर्थ होता जाता है कि यह सत्य है और वह सत्य नहीं है। कविता की जो यात्रा गहराई की ओर है वही उसे अनेकान्त की ओर लिये जा रही है। कवि यह जान गया है कि कोई भी बात जोर से बोलने के योग्य नहीं है। इसलिए अब वह निश्चित और अनिश्चित, ज्ञात और अज्ञात के सन्धिस्थल पर काम करता है। मनुष्य इतनी बार धोखा खा चुका है कि उसे अब किसी भी ज्ञान पर विश्वास नहीं रहा और सत्य को उसने इतनी बार बदलते देखा है कि वह कहीं भी दुराग्रह-पूर्वक अड़ने को तैयार नहीं है। इसका प्रभाव कविता पर पड़ना स्वाभाविक था। कविता अब सत्य का उद्घोष नहीं, उसके अनुसन्धान का प्रयास है। मैं भी 'उर्वशी' में सिखाने के बदले अनुसन्धान के काम में ज्यादा लगा रहा

हूँ। यह ठीक है कि 'उर्वशी' बहुत-से संचित ज्ञान का कथन बड़े ही उत्साह के साथ करती है, किन्तु वह सब का सब सच है या नहीं, यह बात मुझे भी मालूम नहीं है। कविता में एक स्थिति वह भी आती है, जब कवि को अपने अहं का लोप करना पड़ता है अथवा समाधि की स्थिति में देर तक टिके रहने से कवि के अहं का आप से आप लोप हो जाता है। तब तो भूमि रिक्त रह जाती है, वहाँ कहीं से स्रस्त होकर कविता खुद-ब-खुद उतर आती है। 'उर्वशी' में ऐसे कई स्थल हैं। किन्तु उनके बारे में अधिकारपूर्वक बोलना मेरे लिए सम्भव नहीं है, क्योंकि जहाँ-जहाँ ऐसे स्थल आये हैं, मेरा अस्तित्व विलुप्त हो गया है। वहाँ जो है वह कविता है, मैं नहीं हूँ।

इस प्रसंग में 'उर्वशी' को 'कुरुक्षेत्र' से मिलाकर देखने की उत्सुकता स्वाभाविक है। 'कुरुक्षेत्र' में प्रकाश है, 'उर्वशी' में द्वाभा और गोधूलि हैं। 'कुरुक्षेत्र' की वाणी विश्वास की वाणी है, 'उर्वशी' की वाणी संशय और द्विधा से आक्रान्त है। 'कुरुक्षेत्र' में मैं धृष्टतापूर्वक गुरु के पद से स्वयं बोल गया हूँ। 'उर्वशी' की ऊँचाई पर पहुँचकर मुझे ऐसा लगा कि काश, कोई गुरु मिल जाता, तो उससे पूछ लेता कि असली रहस्य क्या है।

अपने स्वप्न, अपनी कल्पना की व्याख्या कवि स्वयं नहीं कर सकता, न यह काम करने की उसे कोशिश करनी चाहिए। कविः करोति काव्यानि रसं जानन्ति पण्डिताः ! फिर भी मैंने यह विवर्जित कार्य इसलिए किया कि मुझे लगा कि इससे आपका किंचित् मनोरंजन हो जाएगा। यह वह युग है, जिसमें माध्यम प्रमुख, सन्देश गौण हो गया है। लोग कविता कम, कविता के बारे में अधिक सुनना चाहते हैं। कवि की जीवनियाँ आज ज्यादा बिकती हैं, उनकी कविताओं की बिक्री कम हो गयी है। कवि की कविता से अधिक महत्त्व अब कवि के साथ की गयी भेंट-वार्ता को दिया जाता है।

जिस सभ्यता में हम जी रहे हैं, वह चौकोर व्यक्तित्ववालों की नहीं, विशेषज्ञों की सभ्यता है। ज्ञान के वृक्ष की डालियाँ अब बढ़कर इतनी स्वतन्त्र हो गयी हैं कि एक डाल पर बसनेवाला पक्षी दूसरी डाल के पक्षी की बोली समझने में असमर्थ है। एक समय ऐसा भी था, जब गेलीलियो, वैज्ञानिक होने पर भी, कविता करते थे और लियोनार्डो द बिंची को, कलाकार होने पर भी, विज्ञान की सारी बातें मालूम थीं। भारत में तो कवि अकसर ज्योतिषी भी हुआ करते थे। ख़ानख़ाना अब्दुर रहीम केवल शायर ही नहीं, ज्योतिषी और सिपहसालार भी थे। लेकिन अब समय ऐसा आ गया है कि भौतिकी के सारे आविष्कार गणित के

फॉरमूलों को समझ सकें। नतीजा यह है कि वैज्ञानिक की बातें सभी वैज्ञानिकों की समझ में नहीं आतीं। इसी प्रकार कवियों की कविताएँ भी कुछ खास-खास कवि ही समझ पाते हैं। यह स्थिति कविता के लिए सामान्य नहीं, अच्छे-खासे सुधी पाठकों के लिए भी दुखदायी हो रही है। काश, कोई ऐसा कवि पैदा होता, जो इलियट और रिल्के के स्वप्नों को तुलसी की सरलता से लिखने का मार्ग निकाल देता।

देवियो और सज्जनो, अन्त में अपने जीवन का एक और भेद बताकर मैं अपना वक्तव्य समाप्त करता हूँ। जिस तरह मैं जवानी भर इक़बाल और रवीन्द्र के बीच झटके खाता रहा, उसी प्रकार मैं जीवन-भर गाँधी और मार्क्स के बीच झटके खाता रहा हूँ। इसीलिएं उजले को लाल से गुणा करने पर जो रंग बनता है, वही रंग मेरी कविता का रंग है। मेरा विश्वास है कि अन्ततोगत्वा यही रंग भारतवर्ष के व्यक्तित्व का भी होगा।

जय वाग्देवी ! जय हिन्द !

–रामधारी सिंह दिनकर

नयी दिल्ली
1.12.1973

अनुक्रम

परिशिष्ट

मंगल-आह्वान

भावों के आवेग प्रबल
मचा रहे उर में हलचल।
कहते, उर के बाँध तोड़
स्वर स्रोतों में बह-बह अनजान,
तृण, तरु, लता, अनिल, जल-थल को
छा लेंगे हम बनकर गान।

पर हूँ विवश, गान से कैसे
जग को हाय! जगाऊँ मैं,
इस तमिस्र युग-बीच ज्योति की
कौन रागिनी गाऊँ मैं?
बाट जोहता हूँ लाचार
आओ, स्वरसम्राट्! उदार।

पल भर को मेरे प्राणों में
ओ विराट् गायक! आओ,
इस वंशी पर रसमय स्वर में
युग-युग के गायन गाओ।

वे गायन, जिन को न आज तक
गाकर सिरा सका जल-थल,
जिन की तान-तान पर आकुल
सिहर-सिहर उठता उडु-दल।

आज सरित का कल-कल, छल-छल
निर्झर का अविरल झर-झर,
पावस की बूँदों की रिम-झिम,
पीले पत्तों का मर्मर,

जलधि-साँस, पक्षी के कलरव,
अनिल-सनन, अलि का गुन-गुन
मेरी वंशी के छिद्रों में
भर दो ये मधु स्वर चुन-चुन।

दो आदेश, फूँक दूँ शृंगी,
उठें प्रभाती-राग महान्,
तीनों काल ध्वनित हों स्वर में,
जागें सुप्त भुवन के प्राण।

गत विभूति, भावी की आशा
ले युगधर्म पुकार उठे,
सिंहों की घन-अन्ध गुहा में
जागृति की हुंकार उठे।

जिन का लुटा सुहाग, हृदय में
उन के दारुण हूक उठे,
चीखूँ यों कि याद कर ऋतुपति
की कोयल रो कूक उठे।

प्रियदर्शन इतिहास कण्ठ में
आज ध्वनित हो काव्य बने,
वर्तमान की चित्रपटी पर—
भूतकाल सम्भाव्य बने।

जहाँ-जहाँ घन-तिमिर हृदय में
भर दो वहाँ विभा प्यारी,

दुर्बल प्राणों की नस-नस में
देव ! फूँक दो चिनगारी।

ऐसा दो वरदान कला को
कुछ भी रहे अजेय नहीं,
रजकण से ले पारिजात तक
कोई रूप अगेय नहीं।

प्रथम खिली जो मधुर ज्योति
कविता वन तमसा-कूलों में,
जो हँसती आ रही युगों से
नभ-दीपों, वनफूलों में;

सूर-सूर, तुलसी-शशि जिस की
विभा यहाँ फैलाते हैं,
जिस के बुझे कणों को पा कवि
अब खद्योत कहाते हैं;

उस की विभा प्रदीप्त करे
मेरे उर का कोना-कोना,
छू दे यदि लेखनी, धूल भी
चमक उठे बनकर सोना।

हिमालय

मेरे नगपति ! मेरे विशाल !

साकार, दिव्य, गौरव विराट्,
पौरुष के पुंजीभूत ज्वाल !
मेरी जननी के हिम-किरीट !,
मेरे भारत के दिव्य भाल !
मेरे नगपति ! मेरे विशाल !

युग-युग अजेय, निर्बन्ध, मुक्त,
युग-युग शुचि, गर्वोन्नत, महान्,
निस्सीम व्योम में तान रहा
युग से किस महिमा का वितान ?

कैसी अखण्ड यह चिर समाधि ?
यतिवर ! कैसा यह अमिट ध्यान ?
तू महाशून्य में खोज रहा
कि जटिल समस्या का निदान ?
उलझन का कैसा विषम जाल ?
मेरे नगपति ! मेरे विशाल !

ओ, मौन तपस्या-लीन यती !
पल भर को तो कर दृगुन्मेष !

रे ज्वालाओं से दग्ध, विकल
है तड़प रहा पद पर स्वदेश।

सुखसिन्धु पंचनद, ब्रह्मपुत्र
गंगा, यमुना की अमिय धार
जिस पुण्य भूमि की ओर बही
तेरी विगलित करुणा उदार,

जिस के द्वारों पर खड़ा क्रान्त
सीमापति ! तू ने की पुकार,
'पद-दलित इसे करना पीछे
पहले ले मेरा सिर उतार।'

उस पुण्य भूमि पर आज तपी !
रे, आन पड़ा संकट कराल,
व्याकुल तेरे सुत तड़प रहे,
डँस रहे चतुर्दिक् विविध व्याल।
मेरे नगपति ! मेरे विशाल !

कितनी मणियाँ लुट गयीं ? मिटा
कितना मेरा वैभव अशेष।
तू ध्यान-मग्न ही रहा, इधर
वीरान हुआ प्यारा स्वदेश।

किन द्रौपदियों के बाल खुले ?
किन-किन कलियों का अन्त हुआ ?
कह हृदय खोल चित्तौर ! यहाँ
कितने दिन ज्वाल-वसन्त हुआ ?

पूछे सिकता-कण से हिमपति,
तेरा वह राजस्थान कहाँ ?

वन-वन स्वतन्त्रता-दीप लिये
फिरनेवाला बलवान् कहाँ ?

तू पूछ अवध से, राम कहाँ ?
वृन्दा ! बोलो, घनश्याम कहाँ ?
ओ मगध ! कहाँ मेरे अशोक ?
वह चन्द्रगुप्त बलधाम कहाँ ?

पैरों पर ही है पड़ी हुई
मिथिला भिखारिनी सुकुमारी,
तू पूछ कहाँ इस ने खोयीं
अपनी अनन्त निधियाँ सारी ?

री कपिलवस्तु ! कह, बुद्धदेव
के वे मंगल उपदेश कहाँ ?
तिब्बत, इरान, जापान, चीन
तक गये हुए सन्देश कहाँ ?

वैशाली के भग्नावशेष से
पूछ लिच्छवी-शान कहाँ ?
ओ री उदास गण्डकी ! बता
विद्यापति कवि के गान कहाँ ?

तू तरुण देश से पूछ अरे,
गूँजा यह कैसा ध्वंस-राग ?
अम्बुधि-अन्तस्तल-बीच छिपी
यह सुलग रही है कौन आग ?

प्राची के प्रांगण-बीच देख,
जल रहा स्वर्ण-युग-अग्नि-ज्वाल
तू सिंहनाद कर जाग तपी !
मेरे नगपति ! मेरे विशाल !

रे, रोक युधिष्ठिर को न यहाँ,
जाने दे उन को स्वर्ग धीर,
पर, फिरा हमें गाण्डीव-गदा,
लौटा दे अर्जुन-भीम वीर।

कह दे शंकर से, आज करें
वे प्रलय-नृत्य फिर एक बार।
सारे भारत में गूँज उठे,
'हर-हर-बम' का फिर महोच्चार।

ले अँगड़ाई, उठ, हिले धरा,
कर निज विराट् स्वर में निनाद,
तू शैलराट्! हुंकार भरे,
फट जाय कुहा भागे प्रमाद!

तू मौन त्याग, कर सिंहनाद,
रे तपी! आज तप का न काल।
नव-युग-शंखध्वनि जगा रही,
तू जाग, जाग, मेरे विशाल!

कविता की पुकार

आज न उडु के नील-कुंज में स्वप्न खोजने जाऊँगी,
आज चमेली में न चन्द्र-किरणों से चित्र बनाऊँगी।
अधरों में मुसकान, न लाली बन कपोल में छाऊँगी,
कवि ! किस्मत पर भी न तुम्हारी आँसू आज बहाऊँगी।
नालन्दा-वैशाली में तुम रुला चुके सौ बार,
धूसर भुवन स्वर्ग–ग्रामों–में कर पायी न विहार।
आज यह राज-वाटिका छोड़, चलो कवि ! वनफूलों की ओर।

चलो, जहाँ निर्जन कानन में वन्य कुसुम मुसकाते हैं,
मलयानिल भूलता, भूलकर जिधर नहीं अलि जाते हैं।
कितने दीप बुझे झाड़ी-झुरमुट में ज्योति पसार ?
चले शून्य में सुरभि छोड़कर कितने कुसुम-कुमार ?
क़ब्र पर मैं कवि ! रोऊँगी, अश्रु-आरती सँजोऊँगी।

विद्युत् छोड़ दीप साजूँगी, महल छोड़ तृण-कुटी-प्रवेश
तुम गाँवों के बनो भिखारी, मैं भिखारिणी का लूँ वेश।

स्वर्णांचला अहा ! खेतों में उतरी सन्ध्या श्याम परी,
रोमन्थन करती गायें आ रहीं रौंदती घास हरी।
घर-घर से उठ रहा धुआँ, जलते चूल्हे बारी-बारी,
चौपालों में कृषक बैठ गाते 'कहँ अटके बनवारी ?'
पनघट से आ रही पीतवसना युवती सुकुमार

किसी भाँति ढोती गागर-यौवन का दुर्वह भार।
बनूँगी मैं कवि ! इस की माँग, कलश, काजल, सिन्दूर सुहाग।

वन-तुलसी की गन्ध लिये हलकी पुरवैया आती है,
मन्दिर की घण्टा-ध्वनि युग-युग का सन्देश सुनाती है।
टिमटिम दीपक के प्रकाश में पढ़ते निज पोथी शिशुगण,
परदेशी की प्रिया बैठ गाती यह विरह-गीत उन्मन,
''भैया ! लिखं दे एक कलम खत मों बालम के जोग,
चारों कोने खेम-कुसल माँझे ठाँ, मोर वियोग।''
दूतिका मैं बन जाऊँगी, सखी ! सुधि उन्हें सुनाऊँगी।

पहन शुक्र का कर्णफूल है दिशा अभी भी मतवाली,
रहते रात रमणियाँ आयीं ले-ले फूलों की डाली।
स्वर्ग-स्रोत, करुणा की धारा, भारत-माँ का पुण्य तरल,
भक्ति-अश्रुधारा-सी निर्मल गंगा बहती है अविरल।
लहर-लहर पर लहराते हैं मधुर प्रभाती-गान,
भुवन स्वर्ग बन रहा, उड़े जाते ऊपर को प्राण,
पुंजारिन की बन कण्ठ-हिलोर, भिगो दूँगी अग-जग के छोर।

कवि ! असाढ़ की इस रिमझिम में धनखेतों में जाने दो,
कृषक-सुन्दरी के स्वर में अटपटे गीत कुछ गाने दो।
दुखियों के केवल उत्सव में इस दम पर्व मनाने दो,
रोऊँगी खलिहानों में, खेतों में तो हर्षाने दो।

मैं बच्चों के संग जरा खेलूँगी दूब-बिछौने पर,
मचलूँगी मैं जरा इन्द्रधनु के रंगीन खिलौने पर।
तितली के पीछे दौड़ूँगी, नाचूँगी, दे-दे ताली,
मैं मकई की सुरभि बनूँगी, पके आम-फल की लाली।

वेणु-कुंज में जुगनू बन मैं इधर-उधर मुसकाऊँगी,
हरसिंगार की कलियाँ बनकर वधुओं पर झर जाऊँगी।

सूखी रोटी खाएगा जब कृषक खेत में धरकर हल,
तब दूँगी मैं तृप्ति उसे बनकर लोटे का गंगाजल।
उस के तन का दिव्य स्वेदकण बनकर गिरती जाऊँगी,
और खेत में उन्हीं कणों से मैं मोती उपजाऊँगी।

शस्य-श्यामता निरख करेगा कृषक अधिक जब अभिलाषा,
तब मैं उसके हृदय-स्रोत में उमड़ूँगी बनकर आशा।
अर्द्धनग्न दम्पति के गृह में मैं झोंका बन आऊँगी।
लज्जित हों न अतिथि-सम्मुख वे, दीपक तुरत बुझाऊँगी।

ऋण-शोधन के लिए दूध-घी बेच-बेच धन जोड़ेंगे,
बूँद-बूँद बेचेंगे, अपने लिए नहीं कुछ छोड़ेंगे।
शिशु मचलेंगे दूध देख, जननी उन को बहलाएगी,
मैं फाड़ूँगी हृदय, लाज से आँख नहीं रो पाएगी।

इतने पर भी धन-पतियों की उन पर होगी मार,
तब मैं बरसूँगी बन बेबस के आँसू सुकुमार।
फटेगा भू का हृदय कठोर, चलो कवि ! वनफूलों की ओर।

परदेशी

माया के मोहक वन की क्या कहूँ कहानी परदेशी ?
भय है, सुनकर हँस दोगे मेरी नादानी परदेशी !
सृजन-बीच संहार छिपा, कैसे बतलाऊँ परदेशी !
सरल कण्ठ से विषम राग कैसे मैं गाऊँ परदेशी ?

एक बात है सत्य कि झर जाते हैं खिलकर फूल यहाँ,
जो अनुकूल, वही बन जाता दुर्दिन में प्रतिकूल यहाँ।
मैत्री के शीतल कानन में छिपा कपट का शूल यहाँ,
कितने काँटों से सेवित है मानवता का मूल यहाँ।
इस उपवन की पगडण्डी पर बचकर जाना परदेशी !
यहाँ मेनका की चितवन पर मत ललचाना परदेशी !

जगती में मादकता देखी, लेकिन, अक्षय तत्त्व नहीं,
आकर्षण में तृप्ति और सुन्दरता में अमरत्व नहीं।
यहाँ प्रेम में मिली विकलता, जीवन में परितोष नहीं,
बाल-युवतियों के आलिंगन में पाया सन्तोष नहीं।
हमें प्रतीक्षा में न तृप्ति की मिली निशानी परदेशी !
माया के मोहक वन की क्या कहूँ कहानी परदेशी ?

महाप्रलय की ओर सभी को इस मरु में चलते देखा,
किस से लिपट जुड़ाता ? सब को ज्वाला में जलते देखा;
अन्तिम बार चिता-दीपक में जीवन को बलते देखा;

चलते समय सिकन्दर-से विजयी को कर मलते देखा।
सब ने देकर प्राण मौत की कीमत जानी परदेशी ?
माया के मोहक वन की क्या कहूँ कहानी परदेशी?

रोते जग की अनित्यता पर सभी विश्व को छोड़ चले,
कुछ तो चढ़े चिता के रथ पर, कुछ कब्रों की ओर चले।
रुके न पल-भर मित्र, पुत्र माता से नाता तोड़ चले,
लैला रोती रही, किन्तु, कितने मजनूँ मुँह मोड़ चले।

जीवन का मधुमय उल्लास,
औ' यौवन का हास-विलास,
रूप-राशि का यह अभिमान,
एक स्वप्न है, स्वप्न अजान।

मिटता लोचन-राग यहाँ पर,
मुरझाती सुन्दरता, प्यारी,
एक-एक कर उजड़ रही है
हरी-भरी कुसुमों की क्यारी।

मैं न रुकूँगा इस भूतल पर
जीवन, यौवन, प्रेम गँवाकर
वायु, उड़ाकर ले चल मुझ को
जहाँ-कहीं इस जग से बाहर।

मरते कोमल वत्स यहाँ, बचती न जवानी परदेशी !
माया के मोहक वन की क्या कहूँ कहानी परदेशी ?

असमय आह्वान

1

समय-असमय का तनिक न ध्यान,
मोहिनी, यह कैसा आह्वान ?

पहन मुक्ता के युग अवतंस,
रत्न-गुम्फित खोले कच-जाल
बजाती मधुर चरण-मंजीर,
आ गयी नभ में रजनी-बाल।

झींगुरों में सुन शिंजन-नाद
मिलन-आकुलता से द्युतिमान,
भेद प्राची का कज्जल-भाल,
बढ़ा ऊपर विधु वेपथुमान।

गया दिन धूलि-धूम के बीच
तुम्हारा करते जय-जयकार,
देखने आया था इस साँझ,
पूर्ण विधु का मादक शृंगार।

एक पल सुधा-वृष्टि के बीच
जुड़ा पाये न क्लान्त मन-प्राण,
कि सहसा गूँज उठा सब ओर
तुम्हारा चिर-परिचित आह्वान।

2

यह कैसा आह्वान !
समय-असमय का तनिक न ध्यान।

झुकी जातीं पलकें निस्पन्द
दिवस के श्रम का लेकर भार,
रहे दृग में क्रम-क्रम से खेल
नये, भोले, लघु स्वप्न-कुमार।

रक्त-कर्दम में दिन-भर फूँक
रजत-शृंगी से भैरव-नाद,
अभी लगता है कितना मधुर
चाँदनी का सुनना संवाद !

दग्ध करती दिन-भर सब अंग
तुम्हारे मरु की जलती धूल;
निशा में ही खिल पाते देवि !
कल्पना के उन्मादक फूल।

अन्य अनुचर सोये निश्चिन्त
शिथिल परियों को करते प्यार;
रात में भी मुझ पर ही पड़ा
द्वार-प्रहरी का दुरुतम भार।

सुलाने आयी गृह-गृह डोल
नींद का सौरभ लिये बतास;
हुए खग नीड़ों में निस्पन्द,
नहीं तब भी मुझ को अवकाश !

ऊँघती इन कलियों को सौंप
कल्पना के मोहक सामान;

पुनः चलना होगा क्या हाय,
तुम्हारा सुन निष्ठुर आह्वान ?

3

यह कैसा आह्वान !
समय-असमय का तनिक न ध्यान।

तुम्हारी भरी सृष्टि के बीच
एक क्या तरल अग्नि ही पेय ?
सुधा-मधु का अक्षय भाण्डार
एक मेरे ही हेतु अदेय ?

'उठो' सुन उठूँ, हुई क्या देवि,
नींद भी अनुचर का अपराध ?
'मरो' सुन मरूँ, नहीं क्या शेष
अभी दो दिन जीने की साध ?

विपिन के फूल-फूल में आज
रही वासन्ती स्वयं पुकार;
अभी भी सुनना होगा देवि !
दुखी धरणी का हाहाकार ?

कर्म क्या एकमात्र वरदान ?
सत्य ही क्या जीवन का श्रेय ?
दग्ध, प्यासी अपनी लघु चाह
मुझे ही रही नहीं क्या गेय ?

मचलता है उडुओं को देख
निकलने जब कोई अरमान;
तभी उठता बज अन्तर-बीच
तुम्हारा यह कठोर आह्वान।

4

यह कैसा आह्वान !
समय-असमय का तनिक न ध्यान।

चाँदनी में छिप किस की ओट
पुष्पधन्वा ने छोड़े तीर ?
बोलने लगी कोकिला मौन,
खोलने लगी हृदय की पीर ?

लताएँ ले द्रुम का अवलम्ब
सजाने लगीं नया शृंगार;
प्रियक-तरु के पुलकित सब अंग
प्रिया का पाकर मधुमय भार।

नहीं यौवन का श्लथ आवेग
स्वयं वसुधा भी सकी सँभाल;
शिराओं का कम्पन ले दिया
सिहरती हरियाली पर डाल।

आज वृन्तों पर बैठे फूल
पहन नूतन, कर्बुर परिधान;
विपिन से लेकर सौरभ-भार
चला उड़ व्योम-ओर पवमान।

किया किस ने यह मधुर स्पर्श ?
विश्व के बदल गये व्यापार।
करेगी उतर व्योम से आज
कल्पना क्या भू पर अभिसार ?

नील कुसुमों के वारिद-बीच
हरे पट का अवगुण्ठन डाल;

स्वामिनी ! यह देखो, है खड़ी
पूर्व-परिचित-सी कोई बाल !

उमड़ता सुषमाओं को देख
आज मेरे दृग में क्यों नीर ?
लगा किसका शर सहसा आन ?
जगी अन्तर में क्यों यह पीर ?

न जाने, किस ने छूकर मर्म ?
जगा दी छवि-दर्शन की चाह;
न जाने चली हृदय को छोड़
खोजने किस को सुरभित आह !

अचानक कौन गया कर क्षुब्ध
न जाने, उर का सिन्धु अथाह ?
जगा किस का यह मादक रोष
रोकने मुझ अजेय की राह ?

न लूँगा आज रजत का शंख,
न गाऊँगा पौरुष का राग,
स्वामिनी ! जलने दो उर-बीच
एक पल तो यह मीठी आग।

तपा लेने दो जी-भर आज
वेदना में प्राणों के गान;
कनक-सा तपकर पीड़ा-बीच
सफल होगा मेरा बलिदान।

चन्द्र-किरणों ने खोले आज;
रुद्ध मेरी आहों के द्वार;
मनाने आ बैठा एकान्त
मधुरता का नूतन त्योहार।

शिथिल दृग में तन्द्रा का भार,
हृदय में छवि का मादक ध्यान;
वेदना का सम्मुख मधु पर्व,
और तब भी दारुण आह्वान !

5

यह कैसा आह्वान !
समय-असमय का तनिक न ध्यान।

चाँदनी की अलकों में गूँथ
छोड़ दूँ क्या अपने अरमान ?
आह ! कर दूँ कलियों में बन्द
मधुर पीड़ाओं का वरदान ?

देवि, कितना कटु सेवा-धर्म !
न अनुचर को निज पर अधिकार
न छिपकर भी कर पाता हाय !
तड़पते अरमानों को प्यार।

हँसो, हिल-डुल वृन्तों के दीप !
हँसो, अम्बर के रत्न अनन्त !
हँसो, हिलमिलकर लता-कदम्ब !
तुम्हें, मंगलमय मधुर वसन्त !

चीरकर मध्य निशा की शान्ति
कोकिले, छेड़ो पंचम तान;
पल्लवों में तुम से भी मधुर
सुला जाता हूँ अपने गान।

भिगोएगी वन के सब अंग
रोर कर जब अब की बरसात,

बजेगा इन्हीं पल्लवों-बीच
विरह मेरा तब सारी रात।

फेंकता हूँ, लो तोड़-मरोड़
अरी निष्ठुरे ! बीन के तार,
उठा चाँदी की उज्ज्वल शंख
फूँकता हूँ भैरव-हुंकार।

नहीं जीते-जी सकता देख
विश्व में झुका तुम्हारा भाल;
वेदना-मधु का भी कर पान
आज उगलूँगा गरल कराल।

सोख लूँ बनकर जिसे अगस्त्य
कहाँ बाधक वह सिन्धु अथाह ?
कहो, खाण्डव-वन वह किस ओर
आज करना है जिसका दाह ?

फोड़ पैठूँ अनन्त पाताल ?
लूट लाऊँ वासव का देश ?
चरण पर रख दूँ तीनों लोक ?
स्वामिनी ! करो शीघ्र आदेश।

किधर होगा अम्बर में दृश्य
देवता का रथ अब की बार ?
शृंग पर चढ़कर जिस के हेतु
करूँ नव स्वागत-मन्त्रोच्चार ?

चाहती हो बुझना यदि आज
होम की शिखा बिना सामान ?
अभय दो, कूद पड़ूँ जय बोल,
पूर्ण कर लूँ अपना बलिदान।

उगे जिस दिन प्राची की ओर
तुम्हारी जय का स्वर्ण विहान,
उगे अंकित नभ पर यह मन्त्र,
'स्वामिनी का असमय आह्वान।'

आलोक धन्वा

ज्योतिर्धर कवि मैं ज्वलित सौर-मण्डल का,
मेरा शिखण्ड अरुणाभ, किरीट अनल का
रथ में प्रकाश के अश्व जुते हैं मेरे
किरणों में उज्ज्वल गीत गुँथे हैं मेरे।

मैं उदय-प्रान्त का सिंह प्रदीप्त विभा से,
केसर मेरे बलते हैं कनक-शिखा से।
ज्योतिर्मयि अन्तःशिखा अरुण है मेरी,
हैं भाव अरुण, कल्पना अरुण है मेरी।

पाया निसर्ग ने मुझे पुण्य के फल-सा,
तम के सिर पर निकला मैं कनक-कमल-सा।
हो उठा दीप्त धरती का कोना-कोना,
जिस को मैंने छू दिया, हुआ वह सोना।

रँग गयी घास पर की शबनम की प्याली,
हो गयी लाल कुहरे की झीनी जाली।
मेरे दृग का आलोक अरुण जब छलका,
बन गयीं घटाएँ बिम्ब उषा-अंचल का।

उदयाचल पर आलोक-शरासन ताने,
आया मैं उज्ज्वल गीत विभा के गाने।

ज्योतिर्धनु की शिंजिनी बजा गाता हूँ,
टंकार-लहर अम्बर में फैलाता हूँ।

किरणों के मुख में विभा बोलती मेरी,
लोहिनी कल्पना उषा खोलती मेरी।
मैं विभापुत्र, जागरण गान है मेरा,
जग को अक्षय आलोक दान है मेरा।

कोदण्ड-कोटि पर स्वर्ग लिये चलता हूँ,
कर-गत दुर्लभ अपवर्ग किये चलता हूँ,
आलोक-विशिख से बेध जगा जन-जन को,
सजता हूँ नूतन शिखा जला जीवन को।

जड़ को उड़ने की पाँख दिये देता हूँ,
चेतन के मन को आँख दिये देता हूँ।
दौड़ा देता हूँ तरल आग नस-नस में,
रहने देता बल को न बुद्धि के बस में।

स्वर को कराल हुंकार बना देता हूँ,
यौवन को भीषण ज्वार बना देता हूँ।
शूरों के दृग अंगार बना देता हूँ,
हिम्मत को ही तलवार बना देता हूँ।

लोहू में देता हूँ वह तेज रवानी,
जूझती पहाड़ों से हो अभय जवानी।
मस्तक में भर अभिमान दिया करता हूँ,
पतनोन्मुख को उत्थान दिया करता हूँ।

म्रियमाण जाति को प्राण दिया करता हूँ।
पीयूष-प्रभा-मय गान दिया करता हूँ।
जो कुछ ज्वलन्त हैं भाव छिपे नर-नर में :
है छिपी विभा उनकी मेरे खर शर में।

किरणें आती हैं समय-वक्ष से कढ़ के,
जाती हैं अपनी राह धनुष पर चढ़ के।
हूँ जगा रहा आलोक अरुण बाणों से
मरघट में जीवन फूँक रहा गानों से।

मैं विभा-पुत्र, जागरण गान है मेरा,
जग को अक्षय आलोक दान है मेरा।

अनल-किरीट

लेना अनल-किरीट भाल पर ओ आशिक होनेवाले !
कालकूट पहले पी लेना, सुधा-बीज बोनेवाले !

1

धरकर चरण विजित शृंगों पर झण्डा वही उड़ाते हैं,
अपनी ही उँगली पर जो ख़ंजर की जंग छुड़ाते हैं।
पड़ी समय से होड़, खींच मत तलवों से काँटे रुककर,
फूँक-फूँक चलती न जवानी चोटों से बचकर, झुककर,
नींद कहाँ उनकी आँखों में जो धुन के मतवाले हैं ?
गति की तृषा और बढ़ती, पड़ते पग में जब छाले हैं।
जागरूक की जय निश्चित है, हार चुके सोनेवाले;
लेना अनल-किरीट भाल पर ओ आशिक होनेवाले !

2

जिन्हें देखकर डोल गयी हिम्मत दिलेर मरदानों की;
उन मौजों पर चली जा रही किश्ती कुछ दीवानों की।
बेफ़िक्री का समाँ कि तूफ़ाँ में भी एक तराना है,
दाँतों उँगली धरे खड़ा अचरज से भरा ज़माना है।
अभय बैठ ज्वालामुखियों पर अपना मन्त्र जगाते हैं,
ये हैं वे, जिनके जादू पानी में आग लगाते हैं।
रूह ज़रा पहचान रखें इन की जादू-टोनेवाले।
लेना अनल-किरीट भाल पर ओ आशिक़ होनेवाले।

3

तीनों लोक चकित सुनते हैं, घर-घर यही कहानी है,
खेल रही नेजों पर चढ़कर रस से भरी जवानी है।
भू सँभले, हो सजग स्वर्ग, यह दोनों की नादानी है
मिट्टी का नूतन पुतला यह अल्हड़ है, अभिमानी है।
अचरज नहीं खींच ईंटें यह सुरपुर को बर्बाद करे।
अचरज नहीं, लूट जन्नत वीरानों को आबाद करे।
तेरी आस लगा बैठे हैं पा-पाकर खोनेवाले।
लेना अनल-किरीट भाल पर ओ आशिक़ होनेवाले।

4

सँभले जग, खिलवाड़ नहीं अच्छा चढ़ते-से पानी से,
याद हिमालय को, भिड़ना कितना है कठिन जवानी से।
ओ महादेश ! बुरा फल है शूरों के शोणित पीने का;
देना होगा तुम्हें एक दिन गिन-गिन मोल पसीने का।
कल होगा इन्साफ़, यहाँ किसने क्या क़िस्मत पायी है।
अभी नींद से जाग रहा युग, यह पहली अँगड़ाई है।
मंज़िल दूर नहीं अपनी, दुख का बोझा ढोने वाले।
लेना अनल-किरीट भाल पर ओ आशिक़ होनेवाले !

दिल्ली

यह कैसी चाँदनी अमा के मलिन तमिस्र गगन में !
कूक रही क्यों नियति व्यंग्य से इस गोधूलि-लगन में ?
मरघट में तू साज रही दिल्ली कैसे शृंगार ?
यह बहार का स्वाँग अरी, इस उजड़े हुए चमन में !

इस उजाड़, निर्जन खँडहर में
छिन्न-भिन्न उजड़े इस घर में,
तुझे रूप सजने की सूझी
मेरे सत्यानाश-प्रहर में।

डाल-डाल पर छेड़ रही कोयल मर्सिया-तराना,
और तुझे सूझा इस दम ही उत्सव हाय, मनाना;
हम धोते हैं घाव इधर सतलज के शीतल जल से,
उधर तुझे भाता है इन पर नमक हाय, छिड़काना !

महल कहाँ ? बस, हमें सहारा
केवल फूस-फाँस, तृणदल का;
अन्न नहीं, अवलम्ब प्राण को
गम, आँसू या गंगाजल का;

यह विहगों का झुण्ड लक्ष्य है
आजीवन वधिकों के फल का;

मरने पर भी हमें कफ़न है
माता शैव्या के अंचल का।

गुल्ची निष्ठुर फेंक रहा कलियों को तोड़ अनल में,
कुछ सागर के पार और कुछ रावी-सतलज-जल में;
हम मिटते जा रहे, न ज्यों अपना कोई भगवान,
यह अलका-छवि-कौन भला देखेगा इस हलचल में ?

बिखरी लट, आँसू छलके हैं,
देख, वन्दिनी है बिलखाती,
अश्रु पोंछने हम जाते हैं,
दिल्ली ! आह ! क़लम रुक जाती।

अरी, विवश हैं, कहो, करें क्या ?
पैरों में ज़ंजीर हाय ! हाथों
में हैं कड़ियाँ कस जातीं।

और कहें क्या ? धरा न धँसती,
हुंकरता न गगन संघाती;
हाय, वन्दिनी माँ के सम्मुख
सुत की निष्ठुर बलि चढ़ जाती।

तड़प-तड़प हम कहो करें क्या ?
'बहै न हाथ, दहै रिसि छाती
अन्तर ही अन्तर घुलते हैं,
'भा कुठार कुण्ठित रिपुघाती।'

अपनी गरदन रेत-रेत असि की तीखी धारों पर
राजहंस बलिदान चढ़ाते माँ के हुंकारों पर।
पगली ! देख, ज़रा कैसी मर मिटने की तैयारी ?
जादू चलेगा न धुन के पक्के इन बनजारों पर।

तू वैभव-मद में इठलाती,
परकीया-सी सैन चलाती,
री ब्रिटेन की दासी ! किस को
इन आँखों पर है ललचाती ?

हम ने देखा यही पाण्डु-वीरों का कीर्ति-प्रसार,
वैभव का सुख-स्वप्न, कला का महास्वप्न-अभिसार,
यहीं कभी अपनी रानी थी, तू ऐसे मत भूल,
अकबर, शाहजहाँ ने जिस का किया स्वयं शृंगार।

तू न ऐंठ मदमाती दिल्ली !
मत फिर यों इतराती दिल्ली !
अविदित नहीं हमें तेरी
कितनी कठोर है छाती दिल्ली !

हाय ! छिनी भूखों की रोटी,
छिना नग्न का अर्द्ध वसन है;
मज़दूरों के कौर छिने हैं,
जिन पर उन का लगा दसन है।

छिनी सजी-साजी वह दिल्ली
अरी ! बहादुरशाह, 'जफ़र' की;
और छिनी गद्दी लखनऊ की
वाजिदअली शाह 'अख़्तर' की।

छिना मुकुट प्यारे 'सिराज' का,
छिना अरी, आलोक नयन का;
नीड़ छिना, बुलबुल फिरती है
वन-वन लिये चंचु में तिनका।

आहें उठीं दीन कृषकों की,
मज़दूरों की तड़प-पुकारें,

अरी ! गरीबों के लोहू पर
खड़ी हुईं तेरी दीवारें।

अंकित है कृषकों के गृह में तेरी निठुर निशानी,
दुखियों की कुटिया रो-रो कहती तेरी मनमानी;
औ, तेरा दृग-मद यह क्या है ? क्या न ख़ून बेकस का ?
बोल, बोल, क्यों लजा रही, ओ कृषक-मेघ की रानी ?

वैभव की दीवानी दिल्ली !
कृषक-मेघ की रानी दिल्ली !
अनाचार, अपमान, व्यंग्य की
चुभती हुई कहानी दिल्ली !

अपने ही पति की समाधि पर
कुलटे ! तू छवि में इतराती;
परदेसी-सँग गलबाँही दे
मन में है फूली न समाती !

दो दिन ही के 'बाल-डान्स' में
नाच हुई बेपानी दिल्ली !
कैसी यह निर्लज्ज नग्नता,
यह कैसी नादानी दिल्ली !

अरी, हया कर, है ज़ईफ़ यह खड़ा कुतुब-मीनार,
इबरत की माँ जामा भी है यहीं अरी ! हुशियार !
इन्हें देखकर भी तो दिल्ली ! आँखें हाय फिरा ले,
गौरव के गुरु रो न पड़ें, हा, घूँघट ज़रा गिरा ले।

अरी, हया कर, हया अभागी !
मत फिर लज्जा को ठुकराती;
चीख न पड़ें क़ब्र में अपनी,
फट न जाय अकबर की छाती।

हूक न उठे कहीं 'दारा' को
कूक न उठे क़ब्र मदमाती!
गौरव के गुरु रो न पड़ें हा,
दिल्ली घूँघट क्यों न गिराती?

बाबर है, औरंग, यहीं है,
मदिरा औ, कुलटा का द्रोही,
बक्सर पर मत भूल, यहीं है
विजयी शेरशाह निर्मोही।

अरी! सँभल, यह क़ब्र न फटकर कहीं बना दे द्वार!
निकल न पड़े क्रोध में लेकर शेरशाह तलवार!
समझाएगा कौन उसे फिर? अरी सँभल नादान!
इस घूँघट पर आज कहीं मच जाय न फिर संहार!

ज़रा गिरा ले घूँघट अपना,
और याद कर वह सुख-सपना,
नूरजहाँ की प्रेम-व्यथा में
दीवाने सलीम का तपना;
गुम्बद पर प्रेमिका कपोती
के पीछे कपोत का उड़ना,
जीवन की आनन्द-घड़ी में
जन्नत की परियों का जुड़ना।

ज़रा याद कर, यहीं नहाती
थी अपनी मुमताज अतर में
तुझ-सी तो सुन्दरी खड़ी
रहती थी पैमाना ले कर में।
सुख, सौरभ, आनन्द बिछे थे
गली, कूच, वन, वीथि, नगर में;
कहती जिसे इन्द्रपुर तू, वह
तो था प्राप्त यहाँ घर-घर में।

आज आँख तेरी बिजली से कौंध-कौंध जाती है !
हमें याद उस स्नेह-दीप की बार-बार आती है !

खिलें फूल, पर, मोह न सकती
हमें अपरिचित छटा निराली;
इन आँखों में घूम रही
अब भी मुरझे गुलाब की लाली।

उठा कसक दिल में लहराती है यमुना का पानी,
पलकें जुगा रहीं बीते वैभव की एक निशानी,
दिल्ली ! तेरे रूप-रंग पर कैसे हृदय फँसेगा ?
बाट जोहती खँडहर में हम कंगालों की रानी !

विपथगा

झन – झन – झन – झन – झन – झनन – झनन,
झन – झन – झन – झन – झन – झनन – झनन।

मेरी पायल झनकार रही तलवारों की झनकारों में,
अपनी आगमनी बजा रही मैं आप क्रुद्ध हुंकारों में,
मैं अहंकार-सी कड़क उठी हँसती विद्युत् की धारों में,
वन काल-हुताशन खेल रही पगली मैं फूट पहाड़ों में,
अँगड़ाई में भूचाल, साँस में लंका के उनचास पवन।
झन-झन-झन-झन-झन-झनन-झनन।

मेरे मस्तक के आतपत्र खर काल-सर्पिणी के शत फन,
मुझ चिर-कुमारिका के ललाट में नित्य नवीन रुधिर-चन्दन,
आँजा करती हूँ चिता-धूम का दृग में अन्ध तिमिर-अंजन,
संहार-लपट का चीर पहन नाचा करती मैं छूम-छनन
झन-झन-झन-झन-झन-झनन-झनन।

पायल की पहली झमक, सृष्टि में कोलाहल छा जाता है,
पड़ते जिस ओर चरण मेरे, भूगोल उधर दब जाता है,
लहराती लपट दिशाओं में, खलभल खगोल अकुलाता है,
परकटे विहग-सा निरवलम्ब गिर स्वर्ग-नरक जल जाता है,
गिरते दहाड़ कर शैल-शृंग मैं जिधर फेरती हूँ चितवन।
झन-झन-झन-झन-झन-झनन-झनन।

रस्सों से कसे जवान पाप-प्रतिकार न जब कर पाते हैं,
बहनों की लुटती लाज देखक़र काँप-काँप रह जाते हैं,
शस्त्रों के भय से जब निरस्त्र आँसू भी नहीं बहाते हैं,
पी अपमानों के गरल-घूँट शासित जब होठ चबाते हैं,
जिस दिन रह जाता क्रोध मौन, मेरा वह भीषण जन्म-लगन।
झन-झन-झन-झन-झन-झनन-झनन।

पौरुष को बेड़ी डाल पाप का अभय रास जब होता है,
ले जगदीश्वर का नाम खड्ग कोई दिल्लीश्वर धोता है,
धन के विलास का बोझ दुखी-दुर्बल दरिद्र जब ढोता है,
दुनिया को भूखों मार भूप जब सुखी महल में सोता है,
सहती सब कुछ मन मार प्रजा, कसमस करता मेरा यौवन।
झन-झन-झन-झन-झन-झनन-झनन।

श्वानों को मिलते दूध-वस्त्र, भूखे बालक अकुलाते हैं,
माँ की हड्डी से चिपक, ठिठुर जाड़ों की रात बिताते हैं,
युवती के लज्जा-वसन बेच जब ब्याज चुकाये जाते हैं,
मालिक जब तेल-फुलेलों पर पानी-सा द्रव्य बहाते हैं,
पापी महलों का अहंकार देता मुझ को तब आमन्त्रण।
झन-झन-झन-झन-झन-झनन-झनन।

डरपोक हुकूमत जुल्मों से लोहा जब नहीं बजाती है,
हिम्मतवाले कुछ कहते हैं, तब जीभ तराशी जाती है,
उलटी चालें ये देख देश में हैरत-सी छा जाती है,
भट्ठी की ओदी आँच छिपी तब और अधिक धुँधुआती है,
सहसा चिंग्घार खड़ी होती दुर्गा मैं करने दस्यु-दलन।
झन-झन-झन-झन-झन-झनन-झनन।

चढ़कर जुनून-सी चलती हूँ मृत्युंजय वीर कुमारों पर,
आतंक फैल जाता क़ानूनी पार्लमेण्ट, सरकारों पर,
'नीरो' के जाते प्राण सूख मेरे कठोर हुंकारों पर,

कर अट्टहास इठलाती हूँ जारों के हाहाकारों पर,
झंझा-सी पकड़ झकोर हिला देती दम्भी के सिंहासन।
झन-झन-झन-झन-झन-झनन-झनन।

मैं निस्तेजों का तेज, युगों के मूक-मौन की बानी हूँ,
दिल-जले शासितों के दिल की मैं जलती हुई कहानी हूँ,
सदियों की जब्ती तोड़ जगी, मैं उस ज्वाला की रानी हूँ,
मैं ज़हर उगलती फिरती हूँ, मैं विष से भरी जवानी हूँ,
भूखी बाघिन की घात क्रूर, आहत भुजंगिनी का दंशन।
झन-झन-झन-झन-झन-झनन-झनन।

जब हुई हुकूमत आँखों पर, जनमी चुपके मैं आहों में,
कोड़ों की खाकर मार पली पीड़ित की दबी कराहों में,
सोने-सी निखर जवान हुई तप कड़े दमन के दाहों में,
ले जान हथेली पर निकली मैं मर मिटने की चाहों में,
मेरे चरणों में खोज रहे भय-कम्पित तीनों लोक शरण।
झन-झन-झन-झन-झन-झनन-झनन।

असि की नोकों से मुकुट जीत अपने सिर उसे सजाती हूँ,
ईश्वर का आसन छीन, कूद मैं आप खड़ी हो जाती हूँ,
थर-थर करते कानून-न्याय इंगित पर जिन्हें नचाती हूँ,
भयभीत पातकी धर्मों से अपने पग मैं धुलवाती हूँ,
सिर झुका घमण्डी सरकारें करतीं मेरा अर्चन-पूजन।
झन-झन-झन-झन-झन-झनन-झनन।

मुझ विपथगामिनी को न ज्ञात, किस रोज़ किधर से आऊँगी,
मिट्टी से किस दिन जाग क्रुद्ध अम्बर में आग लगाऊँगी,
आँखें अपनी कर बन्द देश में जब भूकम्प मचाऊँगी,
किसका टूटेगा शृंग, न जाने, किस का महल गिराऊँगी,
निर्बन्ध, क्रूर, निर्मोह सदा मेरा कराल नर्तन-गर्जन।
झन-झन-झन-झन-झन-झनन-झनन।

अबकी अगस्त की बारी है, पापों के पारावार सजग !
बैठे 'विसूवियस' के मुख पर भोले, अबोध संसार सजग !
रेशों का रक्त कृशानु हुआ, ओ जुल्मी की तलवार सजग !
दुनिया के 'नीरो' सावधान ! दुनिया के पापी 'ज़ार' सजग !
जानें, किस दिन फुंकार उठें पद-दलित काल-सर्पों के फन !
झन-झन-झन-झन-झन-झनन-झनन।

परिचय

सलिल-कण हूँ कि पारावार हूँ मैं ?
स्वयं छाया, स्वयं आधार हूँ मैं ?
बँधा हूँ, स्वप्न है, लघु वृत्त में हूँ,
नहीं तो व्योम का विस्तार हूँ मैं।

समाना चाहती जो बीन-उर में।
विकल वह शून्य की झंकार हूँ मैं।
भटकता, खोजता हूँ ज्योति मन में,
सुना है, ज्योति का आगार हूँ मैं।

जिसे निशि खोजती तारे जलाकर,
उसी का कर रहा अभिसार हूँ मैं।
जनमकर मर चुका सौ बार लेकिन
अगम का पा सका क्या पार हूँ मैं ?

कली की पंखड़ी पर ओस-कण में
रँगीले स्वप्न का संसार हूँ मैं;
मुझे क्या आज ही या कल झड़ूँ मैं ?
सुमन हूँ, एक लघु उपहार हूँ मैं।

जलन हूँ, दर्द हूँ, दिल की कसक हूँ,
किसी का हाय, खोया प्यार हूँ मैं ?

गिरा हूँ भूमि पर नन्दन-विपिन से,
अमर-तरु का सुमन सुकुमार हूँ मैं।

मधुर जीवन हुआ कुछ प्राण ! जब से
लगा ढोने व्यथा का भार हूँ मैं।
रुदन ही एक पथ प्रिय का, इसी से,
पिरोता आँसुओं का हार हूँ मैं।

मुझे क्या गर्व हो अपनी विभा का ?
चिता का धूलि-कण हूँ, क्षार हूँ मैं।
पता मेरा तुम्हें मिट्टी कहेगी,
समा जिस में चुका सौ बार हूँ मैं।

न देखे विश्व, पर, मुझको घृणा से,
मनुज हूँ, सृष्टि का शृंगार हूँ मैं।
पुजारिन ! धूलि से मुझको उठा लो,
तुम्हारे देवता का हार हूँ मैं।

सुनूँ क्या सिन्धु ! मैं गर्जन तुम्हारा ?
स्वयं युग-धर्म का हुंकार हूँ मैं।
कठिन निर्घोष हूँ भीषण अशनि का,
प्रलय-गाण्डीव की टंकार हूँ मैं।

दबी-सी आग हूँ भीषण क्षुधा की,
दलित का मौन हाहाकार हूँ मैं।
सजग संसार, तू निज को सँभाले,
प्रलय का क्षुब्ध पारावार हूँ मैं।

बँधा तूफ़ान हूँ, चलना मना है,
बँधी उद्दाम निर्झर धार हूँ मैं।
कहूँ क्या, कौन हूँ ? क्या आग मेरी ?
बँधी है लेखनी, लाचार हूँ मैं।

गीत-अगीत

गीत, अगीत, कौन सुन्दर है ?

1

गाकर गीत विरह के तटिनी
वेगवती बहती जाती है,
दिल हलका कर लेने को
उपलों से कुछ कहती जाती है।
तट पर एक गुलाब सोचता,
"देते स्वर यदि मुझे विधाता,
अपने पतझर के सपनों का
मैं भी जग को गीत सुनाता।"
गा-गाकर बह रही निर्झरी,
पाटल मूक खड़ा तट पर है।
गीत, अगीत कौन सुन्दर है ?

2

बैठा शुक उस घनी डाल पर
जो खोंते पर छाया देती,
पंख फुला नीचे खोंते में
शुकी बैठ अण्डे है सेती।
गाता शुक जब किरण बसन्ती
छूती अंग पर्ण से छनकर,

किन्तु, शुकी के गीत उमड़कर
रह जाते स्नेह में सनकर।
गूँज रहा शुक का स्वर वन में,
फूला मग्न शुकी का पर है।
गीत, अगीत, कौन सुन्दर है ?

3

दो प्रेमी हैं यहाँ, एक जब
बड़े साँझ आल्हा गाता है,
पहला स्वर उसकी राधा को
घर से यहाँ खींच लाता है।
चोरी-चोरी खड़ी नीम की
छाया में छिपकर सुनती है,
'हुई न क्यों मैं कड़ी गीत की
बिधना', या मन में गुनती है।
वह गाता, पर किसी वेग से
फूल रहा इस का अन्तर है।
गीत, अगीत, कौन सुन्दर है ?

बालिका से वधू

माथे में सेंदुर पर छोटी दो बिन्दी चमचम-सी
पपनी पर आँसू की बूँदें मोती-सी, शबनम-सी।
लदी हुई कलियों से मादक टहनी एक नरम-सी
यौवन की विनती-सी भोली, गुमसुम खड़ी शरम-सी।

पीला चीर, कोर में जिस की चकमक गोटा-जाली,
चली पिया के गाँव उमर के सोलह फूलोंवाली।
पी चुपके आनन्द, उदासी भरे सजल चितवन में,
आँसू में भीगी माया चुपचाप खड़ी आँगन में।

आँखों में दे आँख हेरती हैं उस को जब सखियाँ,
मुस्की आ जाती मुख पर, हँस देतीं रोती अँखियाँ।
पर, समेट लेती शरमाकर बिखरी-सी मुसकान,
मिट्टी उकसाने लगती है अपराधिनी-समान।

भीग रहा मीठा उमंग से दिल का कोना-कोना
भीतर-भीतर हँसी देख लो बाहर-बाहर रोना।

तू वह, जो झुरमुट पर आयी हँसती कनक-कली-सी,
तू वह, जो फूटी शराब की निर्झरिणी पतली-सी।
तू वह, रचकर जिसे प्रकृति ने अपना किया सिंगार,
तू वह, जो धूसर में आयी सबुज रंग की धार।

माँ की ढीठ दुलार ! पिता की ओ लजवन्ती भोली,
ले जाएगी हिया की मणि को अभी पिया की डोली।

कहो, कौन होगी इस घर की तब शीतल उजियारी ?
किसे देख हँस-हँसकर फूलेगी सरसों की क्यारी ?
वृक्ष रीझकर किसे करेंगे पहला फल अर्पण-सा ?
झुकते किस को देख पोखरा चमकेगा दर्पण-सा ?

किस के बाल ओज भर देंगे खुलकर मन्द पवन में ?
पड़ जाएगी जान देखकर किस को चन्द्र-किरन में ?
महँ-महँ कर मंजरी गले से मिल किस को चूमेगी ?
कौन खेत में खड़ी फसल की देवी-सी झूमेगी ?

बनी फिरेगी कौन बोलती प्रतिमा हरियाली की ?
कौन रूह होगी इस धरती फल-फूलोंवाली की ?
हँसकर हृदय पहन लेता जब कठिन प्रेम-ज़ंजीर,
खुलकर तब बजते न सुहागिन, पाँवों के मंजीर।

घड़ी गिनी जाती तब निशि-दिन उँगली की पोरों पर,
प्रिय की याद झूलती है साँसों के हिंडोरों पर
पलती है दिल का रस पीकर सब से प्यारी पीर,
बनती और बिगड़ती रहती पुतली में तस्वीर।

पड़ जाता चस्का जब मोहक प्रेम-सुधा पीने का,
सारा स्वाद बदल जाता है दुनिया में जीने का।
मंगलमय हो पन्थ सुहागिन, यह मेरा वरदान;
हरसिंगार की टहनी-से फूलें तेरे अरमान।

जगे हृदय को शीतल करनेवाली मीठी पीर,
निज को डुबो सके निज में, मन हो इतना गम्भीर।

छाया करती रहे सदा तुझ को सुहाग की छाँह,
सुख-दुख में ग्रीवा के नीचे हो प्रियतम की बाँह।

पल-पल मंगल-लग्न, जिन्दगी के दिन-दिन त्यौहार,
उर का प्रेम फूटकर हो आँचल में उजली धार।

नारी

खिलीं भू पर जब से तुम नारि
कल्पना-सी विधि की अम्लान,
रहे फिर तब से अनु-अनु देवि !
लुब्ध भिक्षुक-से मेरे गान।

तिमिर में ज्योति-कली को देख
सुविकसित, वृन्तहीन, अनमोल;
हुआ व्याकुल सारा संसार,
किया चाहा माया का मोल।

हो उठी प्रतिभा सजग, प्रदीप्त,
तुम्हारी छवि ने मारा बाण;
बोलने लगे स्वप्न निर्जीव,
सिहरने लगे सुकवि के प्राण।

लगे रचने निज उर को तोड़
तुम्हारी प्रतिमा प्रतिमाकार,
नाचने लगी कला चहुँ ओर
भाँवरी दे-दे विविध प्रकार।

ज्ञानियों ने देखा सब ओर
प्रकृति की लीला का विस्तार;

सूर्य शशि, उडु जिन की नख-ज्योति
पुरुष उन चरणों का उपहार।

अगम 'आनन्द'-जलधि में डूब
तृषित 'सत्-चित्' ने पायी पूर्ति;
सृष्टि के नाभि-पद्म पर नारि !
तुम्हारी मिली मधुर रस-मूर्ति।

कुशल विधि-मानस का नवनीत,
एक लघु दिव-सी हो अवतीर्ण,
कल्पना-सी, माया-सी, दिव्य
विभा-सी भू पर हुई विकीर्ण।

दृष्टि तुम ने फेरी जिस ओर
गयी खिल कमल-पंक्ति अम्लान;
हिंस्र मानव के कर से स्रस्त
शिथिल गिर गये धनुष औ' बाण।

हो गया मदिर दृगों को देख
सिंह-विजयी बर्बर लाचार,
रूप के एक तन्तु में नारि,
गया बँध मत्त गयन्द-कुमार।

एक चितवन के शर ने देवि !
सिन्धु को बना दिया परिमेय,
विजित हो दृग-मद से सुकुमारि !
झुका पद-तल पर पुरुष अजेय।

कर्मियों ने देखा जब तुम्हें;
टूटनें लगे शम्भु के चाप।
बेधने चला लक्ष्य गाण्डीव,
पुरुष के खिलने लगे प्रताप।

हृदय निज फ़रहादों ने चीर
बहा दी पय की उज्ज्वल धार
आरती करने को सुकुमारि !
इन्दु को नर ने लिया उतार।

एक इंगित पर दौड़े शूर
कनक-मृग पर होकर हत-ज्ञान,
हुई ऋषियों के तप का मोल
तुम्हारी एक मधुर मुस्कान।

विकल उर को मुरली में फूँक
प्रियक-तरु-छाया में अभिराम,
बजाया हम ने कितनी बार
तुम्हारा मधुमय 'राधा' नाम।

कढ़ीं यमुना से कर तुम स्नान,
पुलिन पर खड़ी हुईं कच खोल,
सिक्त कुन्तल से झरते देवि !
पिये हम ने सीकर अनमोल।

तुम्हारे अधरों का रस प्राण !
वासना-तट पर पिया अधीर;
अरी ओ माँ, हम ने है पिया
तुम्हारे स्तन का उज्ज्वल क्षीर।

पिया शैशव ने रस-पीयूष,
पिया यौवन ने मधु-मकरन्द;
तृषा प्राणों की पर, हे देवि !
एक पल को न सकी हो मन्द।

पुरुष पँखुड़ी को रहा निहार
अयुत जन्मों से छवि पर भूल,

आज तक जान न पाया नारि !
मोहिनी इस माया का मूल।

न छू सकते जिस को हम देवि !
कल्पना वह तुम अगुण, अमेय;
भावना अन्तर की वह गूढ़,
रही जो युग-युग अकथ, अगेय।

तैरतीं स्वप्नों में दिन-रात
मोहिनी छवि-सी तुम अम्लान,
कि जिस के पीछे-पीछे नारि !
रहे फिर मेरे भिक्षुक गान।

रास की मुरली

अभी तक कर पायी न सिंगार,
रास की मुरली उठी पुकार।

1

गयी सहसा किस रस से भीग
बकुल-वन में कोकिल की तान ?
चाँदनी में उमड़ी सब ओर
कहाँ के मद की मधुर उफान ?
गिरा चाहता भूमि पर इन्दु
शिथिलवसना रजनी के संग;
सिहरते पग सकता न सँभाल
कुसुम-कलियों पर स्वयं अनंग !
ठगी-भी रुकी नयन के पास
लिये अंजन उँगली सुकुमार
अचानक लगे नाचने मर्म,
रास की मुरली उठी पुकार।

2

रास की मुरली उठी पुकार।
साँझ तक तो पल गिनती रही,
कहीं तब डूब सका दिनमान;
आँजने जिस क्षण बैठी आँख,

मधुर वेला पहुँची यह आन।
सुहागिनियों में चुनकर एक
मुझे ही भूल गये क्या श्याम ?

बुलाने को न बजाया आज
बाँसुरी में दुखिया का नाम।
बिताऊँ आज रैन किस भाँति ?
पिन्हाऊँ किसे यूथिका-हार ?
धरूँ कैसे घर बैठे धीर ?
रास की मुरली उठी पुकार।

3

रास की मुरली उठी पुकार।
उठी उर में कोमल हिल्लोल
मोहिनी मुरली का सुन नाद,
लगा करने कैसे तो हृदय,
पड़ी, जानें, कैसी कुछ याद।
सकूँगी कैसे स्वयं सँभाल
तरंगित यौवन का रसवाह ?
ग्रन्थि के ढीले कर सब बन्ध
नाचने को आकुल है चाह।
डोलती श्लथ कटि-पट के संग,
खुली रशना करती झनकार,
न दे पायी कंकन में कील,
रास की मुरली उठी पुकार।

4

साज-शृंगार ?
छोड़ दौड़ो सब साज-सिंगार,
रास की मुरली रही पुकार।
अरी भोली मानिनि ! इस रात

विनय-आदर का नहीं विधान,
अनामन्त्रित अर्पण कर देह
पूर्ण करना होगा बलिदान।

आज द्रोही जीवन का पर्व,
नग्न उल्लासों का त्यौहार,
आज केवल भावों का लग्न,
आज निष्फल सारे शृंगार।

अलक्तक-पद का आज न श्रेय,
न कुंकुम की बेंदी अभिराम,
न सोहेगा अधरों में राग,
लोचनों में अंजन घनश्याम।

हृदय का संचित रंग उँड़ेल
सजा नयनों में अनुपम राग,
भीगकर नख-शिख तक सुकुमारि,
आज कर लो निज सुफल सुहाग।

पहनकर केवल मादक रूप
किरण-वसना परियों-सी नग्न,
नीलिमा में हो जाओ बाल,
तारिकामयी प्रकृति-सी मग्न।

यूथिका के ये फूल बिखेर
पुजारिन ! बनो स्वयं उपहार,
पिन्हा बाँहों के मृदुल मृणाल
देवता की ग्रीवा का हार।

खोल बाँहें आलिंगन-हेतु,
खड़ा संगम पर प्राणाधार;

तुम्हें कंकन-कुंकुम का मोह,
और यह मुरली रही पुकार।

5

रास की मुरली रही पुकार।
महालय का यह मंगल-काल,
आज भी लज्जा का व्यवधान ?
तुम्हें तनु पर यदि नहीं प्रतीति,
भेज दो अपने आकुल प्रान।

कहीं हो गया द्विधा में शेष
आज मोहन का मादक रास,
सफल होगा फिर कब सुकुमारि!
तुम्हारे यौवन का मधुमास ?

रही बज आमन्त्रण के राग
श्याम की मुरली नित्य-नवीन,
विकल-सी दौड़-दौड़ प्रतिकाल
सरित हो रही सिन्धु में लीन।

रहा उड़ तज फेनिल अस्तित्व
रूह पल-पल अरूप की ओर
तीव्र होता ज्यों-ज्यों जयनाद,
बढ़ा जाता मुरली का रोर।

सनातन महानन्द में आज
बाँसुरी-कंकन एकाकार,
बहा जा रहा अचेतन विश्व,
रास की मुरली रही पुकार।

निमन्त्रण

तिमिर में स्वर के बाले दीप आज फिर आता है कोई।

‘हवा में कब तक ठहरी हुई
रहेगी जलती हुई मशाल ?
थकी तेरी मुट्ठी यदि वीर,
सकेगा इसको कौन सँभाल ?’
अनल-गिरि पर से मुझे पुकार राग यह गाता है कोई।

हलाहल का दुर्जय विस्फोट,
भरा अंगारों से तूफ़ान,
दहकता-जलता हुआ खगोल,
कड़कता हुआ दीप्त अभियान।
निकट ही कहीं प्रलय का स्वप्न मुझे दिखलाता है कोई।

सुलगती नहीं यज्ञ की आग।
दिशा धूमिल, यजमान अधीर;
पुरोधा-कवि कोई है यहाँ ?
देश को दे ज्वाला के तीर।
धुओं में किसी वह्नि का आज निमन्त्रण लाता है कोई।

आशा का दीपक

वह प्रदीप जो दीख रहा है झिलमिल, दूर नहीं है;
थककर बैठ गये क्या भाई ! मंज़िल दूर नहीं है।

1

चिनगारी बन गयी लहू की बूँद गिरी जो पग से;
चमक रहे, पीछे मुड़ देखो, चरण-चिह्न जगमग से।
शुरू हुई आराध्य-भूमि यह, क्लान्ति नहीं रे राही;
और नहीं तो पाँव लगे हैं क्यों पड़ने डगमग से ?
बाकी होश तभी तक, जब तक जलता तूर नहीं है;
थककर बैठ गये क्या भाई ! मंजिल दूर नहीं है।

2

अपनी हड्डी की मशाल से हृदय चीरते तम का,
सारी रात चले तुम दुख झेलते कुलिश निर्मम का।
एक खेय है शेष, किसी विध पार उसे कर जाओ;
वह देखो, उस पार चमकता है मन्दिर प्रियतम का।
आकर इतना पास फिरे, वह सच्चा शूर नहीं है;
थककर बैठ गये क्या भाई ! मंज़िल दूर नहीं है।

3

दिशा दीप्त हो उठी प्राप्त कर पुण्य-प्रकाश तुम्हारा,
लिखा जा चुका अनल-अक्षरों में इतिहास तुम्हारा।

जिस मिट्टी ने लहू पिया, वह फूल खिलाएगी ही,
अम्बर पर घन बन छाएगा ही उच्छ्वास तुम्हारा।
और अधिक ले जाँच, देवता इतना क्रूर नहीं है।
थककर बैठ गये क्या भाई ! मंज़िल दूर नहीं है।

कलिंग-विजय

1

युद्ध की इति हो गयी; रण-भू श्रमित, सुनसान;
गिरि शिखर पर थम गया है, डूबता दिनमान
देखता यम का भयावह कृत्य,
अन्ध मानव की नियति का नृत्य;

सोचता, इस बन्धु-वध का क्या हुआ परिणाम ?
विश्व को क्या दे गया इतना बड़ा संग्राम ?

युद्ध का परिणाम ?
युद्ध का परिणाम ह्रासत्रास !
युद्ध का परिणाम सत्यानाश !
रुण्ड-मुण्ड-लुठन, निहिंसन, मीच !
युद्ध का परिणाम लोहित कीच !
हो चुका जो कुछ रहा भवितव्य,
यह नहीं नर के लिए कुछ नव्य;
भूमि का प्राचीन यह अभिशाप,
तू गगनचारी ! न कर सन्ताप।
मौन कब के हो चुके रण-तूर्य,
डूब जा तू भी कहीं ओ सूर्य !

छा गया तम, आ गये तारे तिमिर को चीर,
आ गया विधु; किन्तु, क्यों आकृति किये गम्भीर ?

और उस घन-खण्ड ने विधु को लिया क्यों ढाँक ?
फिर गया शशि क्या लजाकर पाप नर के झाँक ?
चाँदनी घन में मिली है छा रही सब ओर,
साँझ को ही दीखता ज्यों हो गया हो भोर।
मौन हैं चारों दिशाएँ, स्तब्ध है आकाश,
श्रव्य जो भी शब्द, वे उठते मरण के पास।

2

शब्द ? यानी घायलों की आह,
घाव के मारे हुओं की क्षीण, करुण कराह,
बह रहा जिसका लहू उस का करुण चीत्कार,
श्वान जिस को नोचते उस की अधीर पुकार।
'घूँट भर पानी, ज़रा पानी' रटन, फिर मौन;
घूँट भर पानी अमृत है, आज देगा कौन ?
बोलते यम के सहोदर श्वान,
बोलते जम्बुक कृतान्त—समान।
मृत्यु-गढ़ पर है खड़ा जयकेतु रेखाकार,
हो गयी हो शान्ति मरघट की यथा साकार।
चल रहा ध्वज के हृदय में द्वन्द्व,
वैजयन्ती है झुकी निस्पन्द।
जा चुके सब लोग फिर आवास,
हतमना कुछ और कुछ सोल्लास।
अंक में घायल मृतक, निश्चेत,
शूर-वीरों को लिटाये रह गया रण-खेत।

और इस सुनसान में निःसंग,
खोजते सच्छान्ति का परिष्वंग,
मूर्तिमय परिताप-से विभ्राट्।
हैं खड़े केवल मगध सम्राट्।
टेक सिर ध्वज का लिये अवलम्ब,
आँख से झर-झर बहाते अम्बु।

भूलकर भूपाल का अहमित्व,
शीश पर वध के लिये दायित्व।
जा चुकी है दृष्टि जग के पार,
आ रहा सम्मुख नया संसार।
चीर वक्षोदेश भीतर पैठ,
दे रहा है सत्य का संवाद,
सुन रहे सम्राट् कोई नाद।
"मन्द मानव ! वासना के भृत्य !
देख ले भर आँख निज दुष्कृत्य।
यह धरा तेरी न थी उपनीत,
शत्रु की त्यों ही नहीं थी क्रीत।
सृष्टि सारी एक प्रभु का राज,
स्वत्व है सब का प्रजा के ब्याज।
मानकर प्रति जीव का अधिकार,
ढो रही धरणी सभी का भार।
एक ही स्तन का पयस कर पान,
जी रहे बलहीन औ' बलवान।
देखने को बिम्ब-रूप अनेक,
किन्तु, दृश्याधार दर्पण एक।
मृत्ति तो बिकती यहाँ बेदाम,
साँस से चलता मनुज का काम।
मृत्तिका हो या कि दीपित स्वर्ण,
साँस पाकर मूर्ति होती पूर्ण।
राज या बल पा अमित अनमोल,
साँस का बढ़ता न किंचित् मोल।
दीनता, दौर्बल्य या अपमान,
त्यों घटा सकते न इस का मान।

तू हुआ सब कुछ, मनुज लेकिन, रहा अब क्या न ?
जो नहीं कुछ बन सका, वह भी मनुज है, मान।
हाय , रे धनलुब्ध जीव कठोर !
हाय रे दारुण ! मुकुटधर भूप लोलुप, चोर!

साजकर इतना बड़ा सामान,
स्वत्व निज सर्वत्र अपना मान।
खड्ग-बल का ले मृषा आधार,
छीनता फिरता मनुज के प्राकृतिक अधिकार।
चरण से प्रभु के नियम को चाप,
तू बना है चाहता भगवान अपना आप।
भौं उठा पाएँ न तेरे सामने बलहीन,
इसलिए ही तो प्रलय यह ! हाय रे हिय-हीन!
शमित करने को स्वमद अति ऊन,
चाहिए तुझ को मनुज का खून।

क्रूरता का साथ ले आख्यान,
जा चुके हैं, जा रहे हैं प्राण।
स्वर्ग में है आज हाहाकार,
चाहता उजड़ा बसा संसार।
भूमि का मानी महीप अशोक,
बाँटता फिरता चतुर्दिक् शोक।
'बाँटता सुतशोक औ' वैधव्य,
बाँटता पशु को मनुज का क्रव्य।

लूटता है गोदियों के लाल,
लूटता सिन्दूर-सज्जित भाल।
यह मनुज-तन में किसी शक्रारि का अवतार
लूट लेता है नगर की सिद्धि, सुख, शृंगार।
शमित करने को स्वमद अति ऊन,
चाहिए उस को मनुज का खून।''

3

आत्म-दर्शन की व्यथा, परिताप, पश्चात्ताप,
डँस रहे सब मिल, उठा है भूप का मन काँप।
स्तब्धता को भेद बारम्बार,
आ रहा है क्षीण हाहाकार।

यह हृदय-द्रावक, करुण वैधव्य का चीत्कार!
यह किसी बूढ़े पिता की भग्न, आर्त पुकार !
यह किसी मृतवत्सला की आह !
आ रही करती हुई दिवदाह !

आ रही है दुर्बलों की हाय,
सूझता है त्राण का नृप को न एक उपाय।
आह की सेना अजेय विराट्,
भाग जा, छिप जा कहीं सम्राट्।

खड्ग से होगी नहीं यह भीत,
तू कभी इस को न सकता जीत।
सामने मन के विरूपाकार,
है खड़ा उल्लंग हो संहार।

षोडशी शुक्लाम्बराएँ आभरण कर दूर,
धूल मलकर धो रही हैं माँग का सिन्दूर।
वीर बेटों की चिताएँ देख ज्वलित समक्ष,
रो रहीं माँएँ हजारों पीटती सिर-वक्ष।

हैं खुले नृप के हृदय के कान;
हैं खुले मन के नयन अम्लान।
सुन रहे हैं विह्वला की आह,
देखते हैं स्पष्ट शव का दाह।
सुन रहे हैं भूप होकर व्यग्र,
रो रहा कैसे कलिंग समग्र !
रो रही हैं वे कि जिन का जल गया शृंगार;
रो रहीं जिन का गया मिट फूलता संसार;
जल गयी उम्मीद, जिन का जल गया है प्यार;
रो रहीं जिन का गया छिन एक ही आधार।

चूड़ियाँ दो एक की प्रतिगृह हुई हैं चूर,
पुछ गया प्रति गेह से दो एक का सिन्दूर।
बुझ गया प्रतिगृह किसी की आँख का आलोक,
इस महा विध्वंस का दायी महीप अशोक।

ध्यान में थे हो रहे आघात,
कान ने सुन ली मगर, यह बात
नाम सुन अपना उसाँसें खींच,
नाक, भौं, आँखें घृणा से मींच,
इस तरह बोले महीपति खिन्न
आप से ज्यों हो गये हों भिन्न–
"विश्व में पापी महीप अशोक,
छीनता है आँख का आलोक।"
देह के दुर्द्धर्ष पशु को मार,
ले चुके हैं देवता अवतार।
निन्द्य लगते पूर्वकृत सब काम,
सुन न सकते आज वे निज नाम।

अश्रु में घुल बह गया कुत्सित, निहीन, विवर्ण,
रह गया है शेष केवल तप्त, निर्मल स्वर्ण।
हूक-सी आकर गयी कोई हृदय़ को तोड़,
ठेस से विष-भाण्ड को कोई गयी है फोड़।

बह गया है अश्रु बनकर कालकूट ज्वलन्त,
जा रहा भरता दया के दूध से वेशन्त।

दूध अन्तर का सरल, अम्लान,
खिल रहा मुख-देश पर द्युतिमान।
किन्तु हैं अब भी झनत्कृत तार,
बोलते हैं भूप बारम्बार–

"हाय रे गर्हित विजय-मद ऊन,
क्या किया मैंने ! बहाया आदमी का खून।"

4

खुल गयी है शुभ्र मन की आँख,
खुल गयी है चेतना की पाँख;
प्राण की अन्तःशिला पर आज पहली बार,
जागकर करुणा उठी है कर मृदुल झनकार।

आँसुओं में गल रहे हैं प्राण,
खिल रहा मन में कमल अम्लान।
गिर गया हतबुद्धि-सा थक कर पुरुष दुर्जेय,
प्राण से निकली अनामय नारि एक अमेय।
अर्धनारीश्वर अशोक महीप;
नर पराजित, नारि सजती है विजय का दीप।

पायलों की सुन मृदुल झनकार,
गिर गयी कर से स्वयं तलवार।
वज्र का उर हो गया दो टूक,
जब उठी कोई हृदय में हूक।

लाल किरणों में यथा हँसता तटी का देश,
एक कोमल ज्ञान से त्यों खिल उठा हृद्देश।
खोल दृग, चारों तरफ अवलोक,
सिर झुका कहने लगे मानी महीप अशोक–

'हे नियन्ता विश्व के कोई अचिन्त्य, अमेय।
ईश या जगदीश कोई शक्ति है अज्ञेय !

हों नहीं क्षन्तव्य जो मेरे विगर्हित पाप,
दो वचन, अक्षय रहे यह ग्लानि, यह परिताप।

प्राण में बल दो, रखूँ निज को सदैव सँभाल,
देव, गर्वस्फीत हो ऊँचा उठे मत भाल।

शत्रु हो कोई नहीं, हो आत्मवत् संसार,
पुत्र-सा पशु-पक्षियों को भी सकूँ कर प्यार।

मिट नहीं जाए किसी का चरण-चिह्न पुनीत,
राह में भी मैं चलूँ पग-पग सजग, सम्भीत।

हो नहीं मुझ को किसी पर रोष,
धर्म का गूँजे जगत् में घोष।

बुद्ध की जय! धर्म की जय! संघ का जय-गान
आ बसें मुझ में तथागत मारजित् भगवान !'
देवता को सौंपकर सर्वस्व,
भूप मन ही मन गये हो निःस्व।

5

और तब उन्मादिनी सोल्लास,
रक्त पर बहती विजय आयी वरण को पास।
संग लेकर ब्याह का उपहार,
रक्त-कर्दम के कमल का हार।
पर, डिगे तिल-भर न वीर महीप;
थी जला करुणा, चुकी तब तक विजय का दीप।

युधिष्ठिर का विलाप

आयी हुई मृत्यु से कहा अजेय भीष्म ने कि
'योग नहीं जाने का अभी है, इसे जानकर,
रुकी रहो पास कहीं', और स्वयं लेट गये।
बाणों का शयन, बाण का ही उपधान कर।
व्यास कहते हैं, रहे यों ही वे पड़े विमुक्त,
काल के करों से छीन मुष्टि-गत प्राण कर,
और पन्थ जोहती विनीत कहीं आसपास
हाथ जोड़ मृत्यु रही खड़ी शास्ति मानकर।
शृंग चढ़ जीवन के आर-पार हेरते-से
योगलीन लेटे थे पितामह गम्भीर-से।
देखा धर्मराज ने, विभा प्रसन्न फैल रही
श्वेत शिरोरुह, शर-ग्रथित शरीर से।
करते प्रणाम, छूते सिर से पवित्र पद,
उँगली को धोते हुए लोचन के नीर से,
'हाय पितामह, महाभारत विफल हुआ',
चीख उठे धर्मराज व्याकुल, अधीर-से।
'वीर-गति पाकर सुयोधन चला गया है,
छोड़ मेरे सामने अशेष ध्वंस का प्रसार;
छोड़ मेरे हाथ में शरीर निज प्राणहीन,
व्योम में बजाता जय-दुन्दुभि-सा बारबार।
और यह मृतक शरीर जो बचा है शेष,
चुप-चुप, मानो, पूछता है मुझ से पुकार–

'विजय का एक उपहार मैं बचा हूँ, बोलो,
जीत किस की है और किस की हुई है हार ?'

'हाय, पितामह, हार किस की हुई है यह ?
ध्वंस-अवशेष पर सिर धुनता है कौन ?
कौन भस्मराशि में विफल सुख ढूँढ़ता है ?
लपटों से मुकुट का पट बुनता है कौन ?
और बैठ मानव की रक्त-सरिता के तीर
नियति के व्यंग्य-भरे अर्थ गुनता है कौन ?
कौन देखता है शवदाह बन्धु-बान्धवों का ?
उत्तरा का करुण विलाप सुनता है कौन ?

'जानता कहीं जो परिणाम महाभारत का,
तन-बल छोड़ मैं मनोबल से लड़ता;
तप से, सहिष्णुता से, त्याग से सुयोधन को
जीत, नयी नींव इतिहास की मैं धरता।
और कहीं वज्र गलता न मेरी आह से जो,
मेरे ताप से नहीं सुयोधन सुधरता;
तो भी हाय, यह रक्त-पात नहीं करता मैं,
भाइयों के संग कहीं भीख माँग मरता।

'किन्तु, हाय, जिस दिन बोया गया युद्ध-बीज,
साथ दिया मेरा नहीं, मेरे दिव्य ज्ञान ने;
उलट दी मति मेरी भीम की गदा ने और
पार्थ के शरासन ने, अपनी कृपाण ने;
और जब अर्जुन को मोह हुआ रण-बीच,
बुझती शिखा में दिया घृत भगवान ने।
सब की सुबुद्धि पितामह, हाय, मारी गयी,
सब को विनष्ट किया एक अभिमान ने।

'कृष्ण कहते हैं, युद्ध, अनघ है, किन्तु मेरे
प्राण जलते हैं पल-पल परिताप से;
लगता मुझे है, क्यों मनुष्य बच पाता नहीं
दह्यमान इस पुराचीन अभिशाप से !
और महाभारत की बात क्या ? गिराये गये
जहाँ छल-छद्म से वरेण्य वीर आप-से,
अभिमन्यु-वध औ' सुयोधन का वध हाय,
हम में बचा है यहाँ कौन, किस पाप से ?

'एक ओर सत्यमयी गीता भगवान की है,
एक ओर जीवन की विरति प्रबुद्ध है;
जानता हूँ, लड़ना पड़ा था हो विवश, किन्तु,
लोहू-सनी जीत मुझे दीखती अशुद्ध है;
ध्वंसजन्य सुख ? याकि, साश्रु-दुख शान्तिजन्य ?
ज्ञात नहीं, कौन बात नीति के विरुद्ध है;
जानता नहीं मैं कुरुक्षेत्र में खिला है पुण्य,
या महान् पाप यहाँ फूटा बन युद्ध है।

'सुलभ हुआ है जो किरीट कुरुवंशियों का,
उस में प्रचण्ड कोई दाहक अनल है;
अभिषेक से क्या पाप मन का धुलेगा कभी ?
पापियों के हित तीर्थ-वारि हलाहल है;
विजय कराल नागिनी-सी डँसती है मुझे
इस से न जूझने को मेरे पास बल है;
ग्रहण करूँ मैं कैसे ? बार-बार सोचता हूँ,
राजसुख लोहू-भरी कीच का कमल है।

'बालहीना माता की पुकार कभी आती, और
आता कभी आर्तनाद पितृहीन बाल का;
आँख पड़ती है जहाँ हाय, वहीं देखता हूँ
सेंदुर पुछा हुआ सुहागिनी के भाल का;

बाहर से भाग कक्ष में जो छिपता हूँ कभी
तो भी सुनता हूँ अट्टहास क्रूर काल का;
और सोते-जागते में चौंक उठता हूँ, मानो,
शोणित पुकारता हो अर्जुन के लाल का।

'जिस दिन समर की अग्नि बुझ शान्त हुई,
एक आग तब से ही जलती है मन में,
हाय, पितामह, किसी भाँति नहीं देखता हूँ
मुँह दिखलाने योग्य निज को भुवन में,
ऐसा लगता है, लोग देखते घृणा से मुझे,
धिक् सुनता हूँ अपने पै कण-कण में।
मानव को देख आँखें आप झुक जातीं, मन
चाहता अकेला कहीं भाग जाऊँ वन में।

'करूँ आत्मघात तो कलंक और घोर होगा,
नगर को छोड़, अतएव, वन जाऊँगा;
पशु-खग भी न देख पाएँ जहाँ, छिप किसी,
कन्दरा में बैठ, अश्रु खुल के बहाऊँगा;
जानता हूँ, पाप न धुलेगा वनवास से भी,
छिपा तो रहूँगा, औ' दुःख तो भुलाऊँगा ?
व्यंग्य से बिंधेगा वहाँ जर्जर हृदय तो नहीं,
वन में कहीं तो धर्मराज न कहाऊँगा।'

अभिनव मनुष्य

है बहुत बरसी धरित्री पर अमृत की धार,
पर, नहीं अब तक सुशीतल हो सका संसार।
भोग-लिप्सा आज भी लहरा रही उद्दाम,
बह रही असहाय नर की भावना निष्काम;

भीष्म हों अथवा युधिष्ठिर, या कि हों भगवान,
बुद्ध हों कि अशोक, गाँधी हों कि ईसु महान्;
सिर झुका सब को, सभी को श्रेष्ठ निज से मान,
मात्र वाचिक ही उन्हें देता हुआ सम्मान,
दग्ध कर पर को, स्वयं भो भोगता दुख-दाह
जा रहा मानव चला अब भी पुरानी राह।

अपहरण शोषण वही, कुत्सित वही अभियान,
खोजना चढ़ दूसरों के भस्म पर उत्थान;
शील से सुलझा न सकना आपसी व्यवहार,
दौड़ना रह-रह उठा उन्माद की तलवार,
द्रोह से अब भी वही अनुराग,
प्राण में अब भी वही फुंकार भरता नाग।

पूर्वयुग-सा आज का जीवन नहीं लाचार,
आ चुका है दूर द्वापर से बहुत संसार;

यह समय विज्ञान का, सब भाँति पूर्ण, समर्थ;
खुल गये हैं गूढ़ संसृति के अमित गुरु अर्थ।
चीरता तम को, सँभाले बुद्धि की पतवार,
आ गया है ज्योति की नव भूमि में संसार।

आज की दुनिया विचित्र, नवीन,
प्रकृति पर सर्वत्र है विजयी पुरुष आसीन।
हैं बँधे नर के करों में वारि, विद्युत्, भाप,
हुक्म पर चढ़ता-उतरता है पवन का ताप।
हैं नहीं बाकी कहीं व्यवधान,
लाँघ सकता नर सरित्, गिरि सिन्धु, एक समान।

शीश पर आदेश कर अवधार्य,
प्रकृति के सब तत्त्व करते हैं मनुज के कार्य।
मानते हैं हुक्म मानव का महा वरुणेश,
और करता शब्दगुण अम्बर वहन सन्देश।
नव्य नर की मुष्टि में विकराल,
हैं सिमटते जा रहे प्रत्येक क्षण दिक्काल।

यह प्रगति निस्सीम ! नर का यह अपूर्व विकास !
चरण-तल भूगोल ! मुट्ठी में निखिल आकाश !

किन्तु है बढ़ता गया मस्तिष्क ही निःशेष,
छूट कर पीछे गया है रह हृदय का देश;
नर मनाता नित्य नूतन बुद्धि का त्योहार,
प्राण में करते दुखी हो देवता चीत्कार।

चाहिए उन को न केवल ज्ञान,
देवता हैं माँगते कुछ स्नेह, कुछ बलिदान;
मोम-सी कोई मुलायम चीज,
ताप पाकर जो उठे मन में पसीज-पसीज;

प्राण के झुलसे विपिन में फूल कुछ सुकुमार;
ज्ञान के मरु में सुकोमल भावना की धार;
चाँदनी की रागिनी, कुछ भोर की मुसकान;
नींद में भूली हुई बहती नदी का गान;
रंग में घुलता हुआ खिलती-कली का राज;
पत्तियों पर गूँजती कुछ ओस की आवाज;
आँसुओं में दर्द की गलती हुई तस्वीर;
फूल की, रस में बसी, भीगी हुई जंजीर।
धूम, कोलाहल, थकावट धूल के उस पार;
शीत जल से पूर्ण कोई मन्दगामी धार;
वृक्ष के नीचे जहाँ मन को मिले विश्राम,
आदमी काटे जहाँ कुछ छुट्टियाँ, कुछ शाम,
कर्म-संकुल लोक-जीवन से समय कुछ छीन,
हो जहाँ पर बैठ नर कुछ पल स्वयं में लीन—
फूल-सा एकान्त में उर खोलने के हेतु,
शाम को दिन की कमाई तोलने के हेतु।

ले चुकी सुख-भाग समुचित से अधिक है देह,
देवता हैं माँगते मन के लिए लघु गेह।

हाय रे मानव, नियति का दास !
हाय रे मनुपुत्र, अपना आप ही उपहास !
प्रकृति की प्रच्छन्नता को जीत,
सिन्धु से आकाश तक सब को किये भयभीत;
सृष्टि को निज बुद्धि से करता हुआ परिमेय,
चीरता परमाणु की सत्ता असीम, अजेय,
बुद्धि के पवमान में उड़ता हुआ असहाय,
जा रहा तू किस दिशा की ओर को निरुपाय ?

लक्ष्य क्या ? उद्देश्य क्या ? क्या अर्थ ?
यह नहीं यदि ज्ञात तो विज्ञान का श्रम व्यर्थ ?

सुन रहा आकाश चढ़ ग्रह-तारकों का नाद;
एक छोटी बात ही पड़ती न तुझको याद ?
एक छोटी, एक सीधी बात,
विश्व में छायी हुई है वासना की रात।
वासना की यामिनी, जिस के तिमिर से हार,
हो रहा नर भ्रान्त अपना आप ही आहार;
बुद्धि में नभ की सुरभि, तन में रुधिर की कीच,
यह वचन से देवता, पर, कर्म से पशु नीच।

यह मनुज,
जिस का गगन में जा रहा है यान,
काँपते जिस के करा को देखकर परमाणु;
खोलकर अपना हृदय गिरि, सिन्धु, भू, आकाश,
हैं सुना जिस को चुके निज गुह्यतम इतिहास।
खुल गये परदे, रहा अब क्या यहाँ अज्ञेय ?
किन्तु नर को चाहिए नित विघ्न कुछ दुर्जेय;
सोचने को और करने को नया संघर्ष,
नव्य जय का क्षेत्र, पाने को नया उत्कर्ष।

पर, धरा सुपरीक्षिता, विश्लिष्ट, स्वाद-विहीन,
यह पढ़ी पोथी न दे सकती प्रवेग नवीन।
एक लघु हस्तामलक यह भूमि-मण्डल गोल,
मानवों ने पढ़ लिये सब पृष्ठ जिस के खोल।
किन्तु, नर-प्रज्ञा सदा गतिशालिनी, उद्दाम,
ले नहीं सकती कहीं रुक एक पल विश्राम।
यह परीक्षित भूमि, यह पोथी पठित, प्राचीन,
सोचने को दे उसे अब बात कौन नवीन ?
यह लघुग्रह भूमिमण्डल, व्योम यह संकीर्ण,
चाहिए नर को नया कुछ और जग विस्तीर्ण।

घुट रही नर-बुद्धि की है साँस;
चाहती वह कुछ बड़ा जग, कुछ बड़ा आकाश।

यह मनुज, जिस के लिए लघु हो रहा भूगोल,
अपर ग्रह-जय की तृषा जिस में उठी है बोल।
यह मनुज विज्ञान में निष्णात,
जो करेगा, स्यात् मंगल और विधु से बात।

यह मनुज, ब्रह्माण्ड का सब से सुरम्य प्रकाश,
कुछ छिपा सकते न जिस से भूमि या आकाश।
यह मनुज, जिस की शिखा उद्दाम,
कर रहे जिस को चराचर भक्तियुक्त प्रणाम।
यह मनुज, जो सृष्टि का शृंगार,
ज्ञान का, विज्ञान का, आलोक का आगार।

पर, सको सुन तो सुनो, मंगल जगत् के लोग !
तुम्हें छूने को रहा जो जीव कर उद्योग—
वह अभी पशु है; निरा पशु, हिंस्र, रक्त-पिपासु,
बुद्धि उस की दानवी है स्थूल की जिज्ञासु,
कड़कता उस में किसी का जब कभी अभिमान,
फूँकने लगते सभी, हो मत्त, मृत्यु-विषाण।

यह मनुज ज्ञानी, शृगालों, कुक्कुरों से हीन—
हो, किया करता अनेकों क्रूर कर्म मलीन।
देह ही लड़ती नहीं, हैं जूझते मन-प्राण,
साथ होते ध्वंस में इस के कला-विज्ञान।
इस मनुज के हाथ से विज्ञान के भी फूल,
वज्र होकर छूटते शुभ धर्म अपना भूल।

यह मनुज, जो ज्ञान का आगार !
यह मनुज, जो सृष्टि का शृंगार।
नाम सुन भूलो नहीं, सोचो-विचारो कृत्य,
यह मनुज, संहार-सेवी, वासना का भृत्य।
छद्म इस की कल्पना, पाषण्ड इस का ज्ञान,
यह मनुष्य, मनुष्यता का घोरतम अपमान।

'व्योम से पाताल तक सब कुछ इसे है ज्ञेय'
पर, न यह परिचय मनुज का, यह न उस का श्रेय।
श्रेय उस का, बुद्धि पर चैतन्य उर की जीत;
श्रेय मानव की असीमित मानवों से प्रीत;
एक नर से दूसरे के बीच का व्यवधान
तोड़ दे जो, बस, वही ज्ञानी, वही विद्वान्
और मानव भी वही।

जो जीव बुद्धि अधीर
तोड़ता अणु ही, न इस व्यवधान का प्राचीर;
वह नहीं मानव; मनुज से उच्च, लघु या भिन्न,
चित्र-प्राणी है किसी अज्ञात ग्रह का छिन्न।
स्यात्, मंगल या शनिश्चर लोक का अवदान,
अजनबी करता सदा अपने ग्रहों का ध्यान।

रसवती भू के मनुज का श्रेय,
यह नहीं विज्ञान, विद्या-बुद्धि यह आग्नेय;
विश्व-दाहक, मृत्यु-वाहक, सृष्टि का सन्ताप,
भ्रान्त पथ पर अन्ध बढ़ते ज्ञान का अभिशाप।
भ्रमित प्रज्ञा का कुतुक यह इन्द्रजाल विचित्र,
श्रेय मानव के न आविष्कार ये अपवित्र।

सावधान, मनुष्य ! यदि विज्ञान है तलवार,
तो इसे दे फेंक, तजकर मोह, स्मृति के पार।
हो चुका है सिद्ध, है तू शिशु अभी अज्ञान;
फूल-काँटा की तुझे कुछ भी नहीं पहचान,
खेल सकता तू नहीं ले हाथ में तलवार,
काट लेगा अंग, तीखी है बड़ी यह धार।

रसवती भू के मनुज का श्रेय,
यह नहीं विज्ञान कटु, आग्नेय।

श्रेय उस का प्राण में बहती प्रणय की वायु,
मानवों के हेतु अर्पित मानवों की आयु।
श्रेय उस का, आँसुओं की धार,
श्रेय उस का भग्न वीणा की अधीर पुकार।
दिव्य भावों के जगत् में जागरण का गान
मानवों का श्रेय आत्मा का किरण-अभियान।
यजन, अर्पण, आत्मसुख का त्याग,
श्रेय मानव का तपस्या की दहकती आग।

बुद्धि-मन्थन से विनिर्गत श्रेय वह नवनीत
जो करे नर के हृदय को स्निग्ध, सौम्य, पुनीत।
श्रेय वह विज्ञान का वरदान,
हो सुलभ सब को सहज जिस का रुचिर अवदान।
श्रेय वह नर-बुद्धि का शिवरूप आविष्कार,
ढो सके जिस से प्रकृति सब के सुखों का भार।
मनुज के श्रम के अपव्यय की प्रथा रुक जाय,
सुख-समृद्धि-विधान में नर के प्रकृति झुक जाय।
श्रेय होगा मनुज का समता-विधायक ज्ञान,
स्नेह-सिंचित न्याय पर नव विश्व का निर्माण।
एक नर में अन्य का निःशंक, दृढ़ विश्वास,
धर्मदीप्त मनुष्य का उज्ज्वल नया इतिहास।
समर, शोषण, ह्रास की विरुदावली से हीन,
पृष्ठ जिस का एक भी होगा न दग्ध, मलीन।
मनुज का इतिहास जो होगा सुधामय कोष,
छलकता होगा सभी नर का जहाँ सन्तोष।
युद्ध की ज्वर-भीति से हो मुक्त,
जब कि होगी, सत्य ही, वसुधा सुधा से युक्त।

श्रेय होगा सुष्ठु-विकसित मनुज का वह काल,
जब नहीं होगी धरा नर के रुधिर से लाल।
श्रेय होगा धर्म का आलोक वह निर्बन्ध,
मनुज जोड़ेगा मनुज से जब उचित सम्बन्ध।

कर्ण-कृष्ण-संवाद

वर्षों तक वन में घूम-घूम,
बाधा-विघ्नों को चूम-चूम,
सह धूप-घाम, पानी-पत्थर,
पाण्डव आये कुछ और निखर
सौभाग्य न सब दिन सोता है,
देखें, आगे क्या होता है।

मैत्री की राह बताने को,
सब को सुमार्ग पर लाने को,
दुर्योधन को समझाने को,
भीषण विध्वंस बचाने को,
भगवान हस्तिनापुर आये,
पाण्डव का सन्देशा लाये।

दो न्याय अगर तो आधा दो,
पर इस में भी यदि बाधा हो,
तो दे दो केवल पाँच ग्राम,
रखो अपनी धरती तमाम।
हम वही खुशी से खाएँगे,
परिजन पर असि न उठाएँगे।

दुर्योधन वह भी दे न सका,
आशिष समाज की ले न सका,

उलटे, हरि को बाँधने चला,
जो था असाध्य, साधने चला।
जब नाश मनुज पर छाता है,
पहले विवेक मर जाता है।

हरि ने भीषण हुंकार किया,
अपना स्वरूप-विस्तार किया,
डगमग-डगमग दिग्गज डोले,
भगवान कुपित होकर बोले—
जंजीर बढ़ाकर साध मुझे,
हाँ-हाँ, दुर्योधन ! बाँध मुझे।

यह देख, गगन मुझ में लय है,
यह देख, पवन मुझ में लय है,
मुझ में विलीन झंकार सकल,
मुझे में लय है संसार सकल।
अमरत्व फूलता है मुझ में,
संहार झूलता है मुझ में।

उदयाचल मेरा दीप्त भाल,
भूमण्डल वक्षस्थल विशाल,
भुज परिधि-बन्ध को घेरे हैं,
मैनाक-मेरु पग मेरे हैं,
दिपते जो ग्रह-नक्षत्र-निकर,
सब हैं मेरे मुख के अन्दर।

दृग हों तो दृश्य अकाण्ड देख,
मुझ में सारा ब्रह्माण्ड देख,
चर-अचर जीव, जग क्षर-अक्षर,
नश्वर मनुष्य, सुरजाति अमर,
शत कोटि सूर्य, शत कोटि चन्द्र,
शत कोटि सरित, सर, सिन्धु मन्द्र;

शत कोटि विष्णु, ब्रह्मा, महेश,
शत कोटि जिष्णु, जलपति, धनेश,
शत कोटि रुद्र, शत कोटि काल,
शत कोटि दण्डधर लोकपाल।
जंज़ीर बढ़ाकर साध इन्हें,
हाँ-हाँ, दुर्योधन ! बाँध इन्हें।

भूलोक, अतल पाताल देख,
गत और अनागत काल देख,
यह देख, जगत् का आदि-सृजन,
यह देख, महाभारत का रण;
मृतकों से पटी हुई भू है,
पहचान, कहाँ इस में तू है।

अम्बर में कुन्तल-जाल देख,
पद के नीचे पाताल देख,
मुट्ठी में तीनों काल देख,
मेरा स्वरूप विकराल देख।
सब जन्म मुझी से पाते हैं,
फिर लौट मुझी में आते हैं।

जिह्वा से कढ़ती ज्वाल सघन,
साँसों में पाता जन्म पवन,
पड़ जाती मेरी दृष्टि जिधर,
हँसने लगती है सृष्टि उधर।
मैं जभी मूँदता हूँ लोचन,
छा जाता चारों ओर मरण।

बाँधने मुझे तो आया है,
ज़ंजीर बड़ी क्या लाया है !
यदि मुझे बाँधना चाहे मन,
पहले तो बाँध अनन्त गगन।

सूने को साध न सकता है,
वह मुझे बाँध कब सकता है ?

हित-वचन नहीं तू ने माना,
मैत्री का मूल्य न पहचाना,
तो ले, मैं भी अब जाता हूँ,
अन्तिम संकल्प सुनाता हूँ।
याचना नहीं, अब रण होगा,
जीवन-जय या कि मरण होगा।

टकराएँगे नक्षत्र निकर,
बरसेगी भू पर वह्नि प्रखर,
फण शेषनाग का डोलेगा,
विकराल काल मुँह खोलेगा।
दुर्योधन ! रण ऐसा होगा,
फिर कभी नहीं जैसा होगा।

भाई पर भाई टूटेंगे,
विष-बाण बूँद-से छूटेंगे,
वायस-शृगाल सुख लूटेंगे,
सौभाग्य मनुज के फूटेंगे।
आखिर तू भूशायी होगा,
हिंसा का पर, दायी होगा।

थी सभा सन्न, सब लोग डरे,
चुप थे या थे बेहोश पड़े।
केवल दो नर न अघाते थे,
धृतराष्ट्र-विदुर सुख पाते थे।
कर जोड़ खड़े प्रमुदित, निर्भय,
दोनों पुकारते थे जय-जय !

2

भगवान सभा को छोड़ चले,
कर के रण-गर्जन घोर चले,
सामने कर्ण सकुचाया-सा,
आ मिला चकित, भरमाया-सा।
हरि बड़े प्रेम से कर धरकर,
ले चढ़े उसे अपने रथ पर।

रथ चला, परस्पर बात चली,
शम-दम की टेढ़ी घात चली।
शीतल हो हरि ने कहा, हाय,
अब शेष नहीं कोई उपाय।
हो विवश हमें धनु धरना है,
क्षत्रिय-समूह को मरना है।

मैं ने कितना कुछ कहा नहीं ?
विष-व्यंग्य कहाँ तक सहा नहीं?
पर, दुर्योधन मतवाला है,
कुछ नहीं समझनेवाला है।
चाहिए उसे बस रण केवल,
सारी धरती कि मरण केवल।

हे वीर ! तुम्हीं बोलो अकाम,
क्या वस्तु बड़ी थी पाँच ग्राम ?
वह भी कौरव को भारी है,
मति गयी मूढ़ की मारी है।
दुर्योधन को बोधूँ कैसे ?
इस रण को अवरोधूँ कैसे ?

सोचो, क्या दृश्य विकट होगा,
रण में जब काल प्रकट होगा ?

बाहर शोणित की तप्त धार,
भीतर विधवाओं की पुकार।
निरशन, विषण्ण बिललाएँगे,
बच्चे अनाथ चिल्लाएँगे।

चिन्ता है, मैं क्या और करूँ ?
शान्ति को छिपा किस ओट धरूँ?
सब राह बन्द मेरे जाने,
हाँ, एक बात यदि तू माने;
तो शान्ति नहीं जल सकती है,
समराग्नि अभी टल सकती है।

पा तुझे धन्य है दुर्योधन,
तू एकमात्र उस का जीवन।
तेरे बल की है आस उसे,
तुझ से जय का विश्वास उसे।
तू संग न उस का छोड़ेगा,
वह क्यों रण से मुख मोड़ेगा ?

क्या अघटनीय घटना कराल ?
तू पृथा-कुक्षि का प्रथम लाल,
बन सूत अनादर सहता है,
कौरव के दल में रहता है,
शर-चाप उठाये आठ प्रहर,
पाण्डव से लड़ने को तत्पर।

माँ का सनेह पाया न कभी,
सामने सत्य आया न कभी,
किस्मत के फेरे में पड़कर,
पा प्रेम बसा दुश्मन के घर।
निज बन्धु मानता है पर को,
कहता है शत्रु सहोदर को।

पर, कौन दोष इस में तेरा ?
अब कहा मान इतना मेरा।
चल होकर संग अभी तेरे,
हैं जहाँ पाँच भ्राता तेरे।
बिछुड़े भाई मिल जाएँगे,
हम मिलकर मोद मनाएँगे।

कुन्ती का तू ही तनय ज्येष्ठ,
बल, बुद्धि में परम श्रेष्ठ।
मस्तक पर मुकुट धरेंगे हम,
तेरा अभिषेक करेंगे हम।
आरती समोद उतारेंगे,
सब मिलकर पाँव पखारेंगे।

पद-त्राण भीम पहनाएगा,
धर्माधिप चँवर डुलाएगा।
पहरे पर पार्थ प्रवर होंगे।
सहदेव-नकुल अनुचर होंगे।
भोजन उत्तरा बनाएगी,
पांचाली पान खिलाएगी।

आहा ! क्या दृश्य सुभग होगा ?
आनन्द-चमत्कृत जग होगा।
सब लोग तुझे पहचानेंगे।
असली स्वरूप में जानेंगे।
खोयी मणि को जब पाएगी,
कुन्ती फूली न समाएगी।

रण अनायास रुक जाएगा,
कुरुराज स्वयं झुक जाएगा।
संसार बड़े सुख में होगा,
कोई न कहीं दुख में होगा।

सब गीत खुशी के गाएँगे,
तेरा सौभाग्य मनाएँगे।

कुरुराज्य समर्पण करता हूँ,
साम्राज्य समर्पण करता हूँ।
यश, मुकुट, मान, सिंहासन ले,
बस, एक भीख मुझ को दे दे।
कौरव को तज रण रोक सखे,
भू का हर भावी शोक सखे।

सुन-सुनकर कर्ण अधीर हुआ,
क्षण एक तनिक गम्भीर हुआ;
फिर कहा, बड़ी यह माया है,
जो कुछ आप ने बताया है।
दिनमणि से सुनकर वही कथा,
मैं भोग चुका हूँ ग्लानि-व्यथा।

जब ध्यान जन्म का धरता हूँ,
उन्मन यह सोचा करता हूँ,
कैसी होगी वह माँ कराल,
निज तन से जो शिशु को निकाल,
धाराओं में धर आती है,
अथवा जीवित दफनाती है ?

सेवती मास दस तक जिस को,
पालती उदर में रख जिस को,
जीवन का अंश खिलाती है,
अन्तर का रुधिर पिलाती है,
आती फिर उस को फेंक कहीं,
नागिन होगी, वह नारि नहीं।

हे कृष्ण ! आप चुप ही रहिए,
इस पर न अधिक कुछ भी कहिए।
सुनना न चाहते तनिक श्रवण,
जिस माँ ने मेरा किया जनन,
वह नहीं नारि कुलपाली थी,
सर्पिणी परम विकराली थी।

पत्थर-समान उस का हिय था,
सुत से समाज बढ़कर प्रिय था,
गोदी में आग लगाकर के,
मेरा कुल-वंश छिपाकर के
दुश्मन का उसने काम किया,
माताओं को बदनाम किया।

माँ का पय भी न पिया मैं ने,
उलटे, अभिशाप लिया मैं ने।
वह तो यशस्विनी बनी रही,
सब की भौं मुझ पर तनी रही।
कन्या वह रही अपरिणीता,
जो कुछ बीता, मुझ पर बीता।

मैं जाति-गोत्र से हीन, दीन,
राजाओं के सम्मुख मलीन,
जब रोज अनादर पाता था
कह शूद्र पुकारा जाता था।
पत्थर की छाती फटी नहीं,
कुन्ती तब भी तो कटी नहीं।

मैं सूत-वंश में पलता था,
अपमान-अनल में जलता था,
सब देख रही थी दृश्य पृथा,
माँ की ममता, पर, हुई वृथा।

छिपकर भी तो सुधि ले न सकी,
छाया अंचल की दे न सकी।

पा पाँच तनय फूली-फूली,
दिन-रात बड़े सुख में भूली,
कुन्ती गौरव में चूर रही,
मुझ पतित पुत्र से दूर रही।
क्या हुआ कि अब अकुलाती है?
किस कारण मुझे बुलाती है ?

क्या पाँच पुत्र हो जाने पर,
सुत के धन-धाम गँवाने पर,
या महानाश के छाने पर,
अथवा मन के घबराने पर,
नारियाँ सदय हो जाती हैं,
बिछुड़े को गले लगाती हैं ?

कुन्ती जिस भय से भरी रही,
तज मुझे, दूर हट खड़ी रही,
वह पाप अभी भी है मुझ में,
वह शाप अभी भी है मुझ में।
क्या हुआ कि वह डर जाएगा ?
कुन्ती को काट न खाएगा ?

सहसा क्या हाल विचित्र हुआ ?
मैं कैसे पुण्य-चरित्र हुआ ?
कुन्ती का क्या चाहता हृदय,
मेरा सुख या पाण्डव की जय ?
यह अभिनन्दन नूतन क्या है ?
केशव ! यह परिवर्तन क्या है ?

मैं हुआ धनुर्धर जब नामी,
सब लोग हुए हित के कामी;
पर, ऐसा भी था एक समय,
जब यह समाज निष्ठुर निर्दय,
किंचित् न स्नेह दर्शाता था,
विष-व्यंग्य सदा बरसाता था।

उस समय सुअंक लगाकर के,
अंचल के तले छिपाकर के,
चुम्बन से कौन मुझे भरकर,
ताड़ना-ताप लेती थी हर ?
राधा को छोड़ भजूँ किस को ?
जननी है वही, तजूँ किस को ?

हे कृष्ण ! ज़रा यह भी सुनिए,
सच है कि झूठ, मन में गुनिए।
धूलों में था मैं पड़ा हुआ,
किस का सनेह पा बड़ा हुआ ?
किस ने मुझ को सम्मान दिया,
नृपता दे महिमावान् किया ?

अपना विकास अवरुद्ध देख,
सारे समाज को क्रुद्ध देख,
भीतर जब टूट चुका था मन,
आ गया अचानक दुर्योधन,
निश्छल, पवित्र अनुराग लिये,
मेरा समस्त सौभाग्य लिये।

कुन्ती ने केवल जन्म दिया,
राधा ने माँ का कर्म किया,
पर, कहते जिसे असल जीवन,
देने आया वह दुर्योधन।

वह नहीं भिन्न माता से है,
बढ़कर सोदर भ्राता से है।

राजा रंक से बनाकर के,
यश, मान, मुकुट पहनाकर के,
बाँहों पर मुझे उठाकर के,
सामने जगत् के लाकर के,
करतब क्या-क्या न किया उसने?
मुझ को नव जन्म दिया उसने।

है ऋणी कर्ण का रोम-रोम,
जानते सत्य यह सूर्य-सोम।
तन, मन, धन, दुर्योधन का है,
यह जीवन दुर्योधन का है।
सुरपुर से भी मुख मोड़ूँगा,
केशव ! मैं उसे न छोड़ूँगा।

सच है, मेरी है आस उसे,
मुझ पर अटूट विश्वास उसे,
हाँ, सच है मेरे ही बल पर
ठाना है उस ने महासमर।
पर, मैं कैसा पापी हूँगा,
दुर्योधन को धोखा दूँगा ?

रह साथ सदा खेला, खाया,
सौभाग्य-सुयश उस से पाया।
अब जब विपत्ति आने को है,
घनघोर प्रलय छाने को है,
तज उसे भाग यदि जाऊँगा,
कायर, कृतघ्न कहलाऊँगा।

मैं भी कुन्ती का एक तनय,
किस को होगा इस का प्रत्यय ?
संसार मुझे धिक्कारेगा,
मन में वह यही विचारेगा,
फिर गया तुरत जब राज मिला,
यह कर्ण बड़ा पापी निकला।

मैं ही न सहूँगा विषम डंक,
अर्जुन को भी होगा कलंक,
सब लोग कहेंगे डरकर ही,
अर्जुन ने अद्भुत नीति गही।
चल चाल कर्ण को फोड़ लिया,
सम्बन्ध अनोखा जोड़ लिया।

कोई न कहीं भी चूकेगा,
सारा जग मुझ पर थूकेगा,
तप, त्याग, शील, जप, याग, दान,
मेरे होंगे मिट्टी समान।
लोभी-लालची कहाऊँगा,
किस को, क्या मुख दिखलाऊँगा?

जो आज आप कह रहे आर्य,
कुन्ती के मुख से कृपाचार्य,
सुन वही हुए लज्जित होते,
हम क्यों रण को सज्जित होते ?
मिलता न कर्ण दुर्योधन को,
पाण्डव न कभी जाते वन को।

लेकिन, नौका तट छोड़ चली,
कुछ पता नहीं, किस ओर चली।
वह बीच नदी की धारा है,
सूझता न कूल-किनारा है।

ले लील भले यह धार मुझे,
लौटना नहीं स्वीकार मुझे।

धर्माधिराज का ज्येष्ठ बनूँ ?
भारत में सब से श्रेष्ठ बनूँ ?
कुल की पोशाक पहनकर के,
सिर उठा चलूँ कुछ तनकर के ?
इस झूठ-मूठ में रस क्या है ?
केशव ! यह सुयश सुयश क्या है?

सिर पर कुलीनता का टीका,
भीतर जीवन का रस फीका,
अपना न नाम जो ले सकते,
परिचय न तेज से दे सकते,
ऐसे भी कुछ नर होते हैं,
कुल को खाते औ' खोते हैं।

विक्रमी पुरुष लेकिन, सिर पर
चलता न छत्र पुरखों का धर,
अपना बल-तेज जगाता है,
सम्मान जगत् से पाता है।
सब उसे देख ललचाते हैं,
कर विविध यत्न अपनाते हैं।

कुल-गोत्र नहीं साधन मेरा,
पुरुषार्थ एक बस धन मेरा,
कुल ने तो मुझ को फेंक दिया,
मैं ने हिम्मत से काम लिया।
अब वंश चकित भरमाया है,
खुद मुझे खोजने आया है।

लेकिन, मैं लौट चलूँगा क्या ?
अपने प्रण से बिचलूँगा क्या ?
रण में कुरुपति का विजय-वरण,
या पार्थ-हाथ कर्ण का मरण।
हे कृष्ण ! यही मति मेरी है,
तीसरी नहीं गति मेरी है।

मैत्री की बड़ी सुखद छाया,
शीतल हो जाती है काया,
धिक्कार-योग्य होगा वह नर,
जो पाकर भी ऐसा तरुवर,
हो अलग खड़ा कटवाता है,
खुद आप नहीं कट जाता है।

जिस नर की बाँह गही मैं ने,
जिस तरु की छाँह गही मैं ने,
उस पर न वार चलने दूँगा;
कैसे कुठार चलने दूँगा ?
जीते जी उसे बचाऊँगा,
या आप स्वयं कट जाऊँगा।

मित्रता बड़ा अनमोल रतन,
कब इसे तोल सकता है धन ?
धरती की तो है क्या बिसात ?
आ जाय अगर वैकुण्ठ हाथ,
उस को भी न्योछावर कर दूँ,
कुरुपति के चरणों पर धर दूँ।

सिर लिये स्कन्ध पर चलता हूँ,
उस दिन के लिए मचलता हूँ,
यदि चले वज्र दुर्योधन पर,
ले लूँ बढ़कर अपने ऊपर।

कटवा दूँ उस के लिए गला,
चाहिए मुझे क्या और भला ?

सम्राट् बनेंगे धर्मराज,
या पाएगा कुरुराज ताज;
लड़ना भर मेरा काम रहा,
दुर्योधन का संग्राम रहा।
मुझ को न कहीं कुछ पाना है,
केवल ऋण मात्र चुकाना है।

कुरुराज्य चाहता मैं कब हूँ ?
साम्राज्य चाहता मैं कब हूँ ?
क्या नहीं आप ने भी जाना ?
मुझ को न आज तक पहचाना?
जीवन का मूल समझता हूँ,
धन को मैं धूल समझता हूँ।

धनराशि जोगना लक्ष्य नहीं,
साम्राज्य भोगना लक्ष्य नहीं,
भुजबल से कर संसार-विजय,
अगणित समृद्धियों का संचय,
दे दिया मित्र दुर्योधन को,
तृष्णा छू भी न सकी मन को।

वैभव-विलास की चाह नहीं,
अपनी कोई परवाह नहीं,
बस, यही चाहता हूँ केवल,
दान की देव-सरिता निर्मल,
करतल से झरती रहे सदा,
निर्धन को भरती रहे सदा।

तुच्छ है, राज्य क्या है केशव ?
पाता नर क्या कर प्राप्त विभव ?
चिन्ता प्रभूत, अत्यल्प हास,
कुछ चाकचिक्य, कुछ क्षण विलास।
पर, वह भी यहीं गँवाना है,
कुछ साथ नहीं ले जाना है।

मुझ-से मनुष्य जो होते हैं,
कंचन का भार न ढोते हैं।
पाते हैं धन बिखराने को,
लाते हैं रतन लुटाने को।
जग से न कभी कुछ लेते हैं,
दान ही हृदय का देते हैं।

प्रासादों के कनकाभ शिखर,
होते कबूतरों के ही घर,
महलों में गरुड़ न होता है,
कंचन पर कभी न सोता है।
बसता वह कहीं पहाड़ों में,
शैलों की फटी दरारों में।

होकर समृद्धि-सुख के अधीन,
मानव होगा नित तपःक्षीण,
सत्ता, किरीट, मणिमय आसन
करते मनुष्य का तेज-हरण।
नर विभव-हेतु ललचाता है,
पर, वही मनुज को खाता है।

चाँदनी, पुष्प-छाया में पल,
नर भले बने सुमधुर, कोमल,
पर अमृत क्लेश का पिये बिना,
आतप, अन्धड़ में जिये बिना,

वह पुरुष नहीं कहला सकता,
विघ्नों को नहीं हिला सकता।

उड़ते जो झंझावातों में,
पीते जो वारि प्रपातों में,
सारा आकाश अयन जिन का,
विषधर भुजंग भोजन जिन का
वे ही फणिबन्ध छुड़ाते हैं,
धरती पर हृदय जुड़ाते हैं।

मैं गरुड़ कृष्ण ! मैं पक्षिराज,
सिर पर न चाहिए मुझे ताज।
दुर्योधन पर है विपद घोर,
सकता न किसी विध उसे छोड़।
रणखेत पाटना है मुझ को,
अहिपाश काटना है मुझ को।

संग्राम-सिन्धु लहराता है,
सामने प्रलय घहराता है,
रह-रहकर भुजा फड़कती है,
बिजली-सी नसें कड़कती हैं।
चाहता तुरत मैं कूद पड़ूँ,
जीतूँ कि समर में डूब मरूँ।

अब देर नहीं कीजै केशव !
अवसेर नहीं कीजै केशव !
धनु की डोरी तन जाने दें,
संग्राम तुरत ठन जाने दें।
ताण्डवी तेज लहराएगा,
संसार ज्योति कुछ पाएगा।

पर, एक विनय है मधुसूदन !
मेरी यह जन्मकथा गोपन,
मत कभी युधिष्ठिर से कहिए,
जैसे हो, इसे दबा रहिए।
वे इसे जान यदि पाएँगे,
सिंहासन को ठुकराएँगे।

साम्राज्य न कभी स्वयं लेंगे,
सारी सम्पत्ति मुझे देंगे,
मैं भी न उसे रख पाऊँगा,
दुर्योधन को दे जाऊँगा।
पाण्डव वंचित रह जाएँगे,
दुख से न छूट वे पाएँगे।

अच्छा, अब चला, प्रणाम आर्य !
हों सिद्ध समर के शीघ्र कार्य।
रण में ही अब दर्शन होगा,
शर से चरण-स्पर्शन होगा।
जय हो, दिनेश नभ में विहरें,
भूतल में दिव्य प्रकाश भरें।

रथ से राधेय उतर आया,
हरि के मन में विस्मय छाया,
बोले कि वीर ! शत बार धन्य,
तुझ-सा न मित्र कोई अनन्य।
तू कुरुपति का ही नहीं प्राण,
नरता का है भूषण महान।

भारत का यह रेशमी नगर

हो गया एक नेता मैं भी ? तो बन्धु, सुनो,
मैं भारत के रेशमी नगर में रहता हूँ,
जनता तो चट्टानों का बोझ सहा करती,
मैं चाँदनियों का बोझ किसी विध सहता हूँ।

दिल्ली फूलों में बसी, ओस-कण से भीगी,
दिल्ली सुहाग है, सुषमा है, रंगीनी है
प्रेमिका-कण्ठ में पड़ी मालती की माला,
दिल्ली सपनों की सेज मधुर रस-भीनी है।

बस, जिधर उठाओ दृष्टि, उधर रेशम केवल,
रेशम पर से क्षण भर को आँख न हटती है,
सच कहा एक भाई ने, दिल्ली में तन से
रेशम से रुखड़ी चीज़ न कोई सटती है।

आखिर हो भी क्यों नहीं ? कि दिल्ली के भीतर
जाने, युग से कितनी सिद्धियाँ समायी हैं।
औ' सब का पहुँचा काल तभी जब से उन की
आँखें रेशम पर बहुत अधिक ललचायी हैं।

रेशम के कोमल तार, क्लान्तियों के धागे,
हैं बँधे उन्हीं से अंग यहाँ आजादी के,

दिल्लीवाले गा रहे बैठ निश्चिन्त, मगन
रेशमी महल में गीत खुरदरी खादी के।

वेतनभोगिनी, विलासमयी यह देवपुरी,
ऊँघती कल्पनाओं से जिस का नाता है,
जिस को इतनी चिन्ता का भी अवकाश नहीं
खाते हैं जो वह अन्न कौन उपजाता है।

उद्यानों का यह नगर, कहीं भी जा देखो,
इस में कुम्हार का चाक न कोई चलता है,
मज़दूर मिलें, पर, मिलता कहीं किसान नहीं,
फूलते फूल, पर, मक्का कहीं न फलता है।

क्या ताना है मोहक वितान मायापुर का,
बस, फूल-फूल, रेशम-रेशम फैलाया है;
लगता है, कोई स्वर्ग खमण्डल से उड़कर,
मदिरा में माता हुआ भूमि पर आया है।

ये, जो फूलों के चीरों में चमचमा रहीं,
मधुमुखी इन्द्रजाया की सहचरियाँ होंगी,
ये, जो यौवन की धूम मचाये फिरती हैं,
भूतल पर भटकी हुई इन्द्रपरियाँ होंगी।

उभरे गुलाब से घटकर कोई फूल नहीं,
नीचे कोई सौन्दर्य न कसी जवानी से,
दिल्ली की सुषमाओं का कौन बखान करे ?
कम नहीं कड़ी कोई भी स्वप्न-कहानी से।

गन्दगी, गरीबी, मैलेपन को दूर रखो,
शुद्धोदन के पहरेवाले चिल्लाते हैं;
है कपिलवस्तु पर फूलों का शृंगार पड़ा,
रथ-समारूढ़ सिद्धार्थ घूमने जाते हैं।

सिद्धार्थ देख रम्यता रोज ही फिर आते;
मन में कुत्सा का भाव नहीं, पर, जगता है;
समझाये उन को कौन, नहीं भारत वैसा
दिल्ली के दर्पण में जैसा वह लगता है ?

भारत धूलों से भरा, आँसुओं से गीला,
भारत अब भी व्याकुल विपत्ति के घेरे में।
दिल्ली में तो है खूब ज्योति की चहल-पहल,
पर, भटक रहा है सारा देश अँधेरे में।

रेशमी कलम से भाग्य-लेख लिखनेवालो,
तुम भी अभाव से कभी ग्रस्त हो रोये हो ?
बीमार किसी बच्चे की दवा जुटाने में,
तुम भी क्या घर भर पेट बाँधकर सोये हो ?

असहाय किसानों की किस्मत को खेतों में,
क्या अनायास जल में बह जाते देखा है ?
'क्या खाएँगे ?' यह सोच निराशा से पागल,
बेचारों को नीरव रह जाते देखा है ?

देखा है ग्रामों की अनेक रम्भाओं को,
जिन की आभा पर धूल अभी तक छायी है?
रेशमी देह पर जिन अभागिनों की अब तक
रेशम क्या ? साड़ी सही नहीं चढ़ पायी है।

पर, तुम नगरों के लाल, अमीरी के पुतले,
क्यों व्यथा भाग्यहीनों की मन में लाओगे ?
जलता हो सारा देश, किन्तु, होकर अधीर
तुम दौड़-दौड़कर क्यों यह आग बुझाओगे ?

चिन्ता हो भी क्यों तुम्हें, गाँव के जलने से,
दिल्ली में तो रोटियाँ नहीं कम होती हैं।

धुलता न अश्रु-बूँदों से आँखों का काजल,
गालों पर की धूलियाँ नहीं नम होती हैं।

जलते हैं तो ये गाँव देश के जला करें,
आराम नयी दिल्ली अपना कब छोड़ेगी ?
या रक्खेगी मरघट में भी रेशमी महल,
या आँधी की खाकर चपेट सब छोड़ेगी।

चल रहे ग्राम-कुंजों में पछिया के झकोर,
दिल्ली, लेकिन, ले रही लहर पुरवाई में।
है विकल देश सारा अभाव के तापों से;
दिल्ली सुख से सोयी है नरम रजाई में।

क्या कुटिल व्यंग्य ! दीनता वेदना से अधीर
आशा से जिन का नाम रात-दिन जपती है,
दिल्ली के वे देवता रोज कहते जाते,
'कुछ और धरो धीरज, किस्मत अब छपती है।'

किस्मतें रोज़ छप रहीं, मगर जलधार कहाँ ?
प्यासी हरियाली सूख रही है खेतों में;
निर्धन का धन पी रहे लोभ के प्रेत छिपे,
पानी विलीन होता जाता है रेतों में।

हिल रहा देश कुत्सा के जिन आघातों से,
वे नाद तुम्हें ही नहीं सुनाई पड़ते हैं ?
निर्माणों के प्रहरियो ! तुम्हें ही चोरों के
काले चेहरे क्या नहीं दिखाई पड़ते हैं ?

तो होश करो, दिल्ली के देवो, होश करो,
सब दिन तो यह मोहिनी न चलनेवाली है;
होती जाती हैं गर्म दिशाओं की साँसें,
मिट्टी फिर कोई आग उगलनेवाली है।

हो रहीं खड़ी सेनाएँ फिर काली-काली
मेघों-से उभरे हुए नये गजराजों की,
फिर नये गरुड़ उड़ने को पाँखें तोल रहे,
फिर झपट झेलनी होगी नूतन बाजों की।

वृद्धता भले बँध रहे रेशमी धागों से,
साबित इन को, पर, नहीं जवानी छोड़ेगी;
जिस के आगे झुक गये सिद्धियों के स्वामी,
उस जादू को कुछ नयी आँधियाँ तोड़ेंगी।

ऐसा टूटेगा मोह, एक दिन के भीतर
इस राग-रंग की पूरी बर्बादी होगी,
जब तक न देश के घर-घर में रेशम होगा,
तब तक दिल्ली के भी तन पर खादी होगी।

नील कुसुम

‘‘है यहाँ तिमिर, आगे भी ऐसा ही तम है,
तुम नील कुसुम के लिए कहाँ तक जाओगे ?
जो गया, आज तक नहीं कभी वह लौट सका,
नादान मर्द ! क्यों अपनी जान गँवाओगे ?

प्रेमिका ? अरे, उन शोख बुतों का क्या कहना !
वे तो यों ही उन्माद जगाया करती हैं,
पुतली से लेतीं बाँध प्राण की डोर प्रथम,
पीछे चुम्बन पर क़ैद लगाया करती हैं।

इन में से किस ने कहा, चाँद से कम लूँगी ?
पर, चाँद तोड़कर कौन मही पर लाया है ?
किस के मन की कल्पना गोद में बैठ सकी ?
किस का जहाज फिर देश लौटकर आया है ?’’

ओ नीतिकार ! तुम झूठ नहीं कहते होगे;
बेकार मगर, पगलों को ज्ञान सिखाना है;
मरने का होगा ख़ौफ़ मौत की छाती में,
जिस को अपनी जिन्दगी ढूँढ़ने जाना है ?

औ’ सुना कहाँ तुम ने कि ज़िन्दगी कहते हैं,
सपनों ने देखा जिसे, उसे पा जाने को ?

इच्छाओं की मूर्तियाँ घूमतीं जो मन में,
उन को उतार मिट्टी पर गले लगाने को ?

ज़िन्दगी, आह ! वह एक झलक रंगीनी की,
नंगी उँगली जिस को न कभी छू पाती है,
हम जभी हाँफते हुए चोटियों पर चढ़ते,
वह खोल पंख चोटियाँ छोड़ उड़ जाती है।

रंगीनी की वह एक झलक, जिस के पीछे
है मची हुई आपा-आपी मस्तानों में,
वह एक दीप जिस के पीछे हैं डूब रहीं
दीवानों की किश्तियाँ कठिन तूफ़ानों में।

डूबती हुई किश्तियाँ ! और यह किलकारी !
ओ नीतिकार ! क्या मौत इसी को कहते हैं ?
है यही ख़ौफ़, जिस से डरकर जीनेवाले
पानी से अपना पाँव समेटे रहते हैं ?

ज़िन्दगी गोद में उठा-उठा हलराती है
आशाओं की भीषिका झेलनेवालों को;
औ' बड़े शौक से मौत पिलाती है जीवन
अपनी छाती से लिपट खेलनेवालों को।

तुम लाशें गिनते रहे खोजनेवालों की,
लेकिन उन की असलियत नहीं पहचान सके;
मुरदों में केवल यही ज़िन्दगीवाले थे—
जो फूल उतारे बिना लौटकर आ न सके।

हो जहाँ कहीं भी नील कुसुम की फुलवारी,
मैं एक फूल तो किसी तरह ले जाऊँगा,
जूड़े में जब तक भेंट नहीं यह बाँध सकूँ,
किस तरह प्राण की मणि को गले लगाऊँगा ?

चाँद और कवि

रात यों कहने लगा मुझ से गगन का चाँद,
आदमी भी क्या अनोखा जीव होता है।
उलझनें अपनी बनाकर आप ही फँसता,
और फिर बेचैन हो जगता न सोता है।

जानता है तू कि मैं कितना पुराना हूँ ?
मैं चुका हूँ देख मनु को जनमते-मरते;
और लाखों बार तुझ-से पागलों को भी
चाँदनी में बैठ स्वप्नों पर सही करते।

आदमी का स्वप्न ? है वह बुलबुला जल का,
आज बनता और कल फिर फूट जाता है;
किन्तु, तो भी धन्य; ठहरा आदमी ही तो !
बुलबुलों से खेलता, कविता बनाता है।

मैं न बोला, किन्तु मेरी रागिनी बोली,
चाँद ! फिर से देख, मुझ को जानता है तू ?
स्वप्न मेरे बुलबुले हैं ? है यही पानी ?
आग को भी क्या नहीं पहचानता है तू ?

मैं न वह जो स्वप्न पर केवल सही करते,
आग में उस को गला लोहा बनाती हूँ;

और उस पर नींव रखती हूँ नये घर की,
इस तरह, दीवार फ़ौलादी उठाती हूँ।

मनु नहीं, मनु-पुत्र है यह सामने, जिस की
कल्पना की जीभ में भी धार होती है;
बाण ही होते विचारों के नहीं केवल,
स्वप्न के भी हाथ में तलवार होती है।

स्वर्ग के सम्राट् को जाकर खबर कर दे,
"रोज़ ही आकाश चढ़ते जा रहे हैं ये;
रोकिए, जैसे बने, इन स्वप्नवालों को,
स्वर्ग की ही ओर बढ़ते जा रहे हैं ये।"

व्याल-विजय

झूमे ज़हर चरण के नीचे, मैं उमंग में गाऊँ,
तान, तान फण व्याल, कि तुझ पर मैं बाँसुरी बजाऊँ।

1

यह बाँसुरी बजी माया के मुकुलित आकुंचन में,
यह बाँसुरी बजी अविनाशी के संवेश गहन में,
अस्तित्वों के अनस्तित्व में, महाशान्ति के तल में,
यह बाँसुरी बजी शून्यासन की समाधि निश्चल में।
कम्पहीन तेरे समुद्र में जीवन-लहर उठाऊँ,
तान, तान फण व्याल, कि तुझ पर मैं बाँसुरी बजाऊँ।

2

अक्षय-वट पर बजी बाँसुरी, गगन, मगन लहराया
दल पर विधि को लिये जलधि में नाभि-कमल उठ आया।
जनमी नव चेतना, सिहरने लगे तत्त्व चलदल-से,
स्वर का ले अवलम्ब भूमि निकली प्लावन के जल से।
अपने आर्द्र-वसन की वसुधा को फिर याद दिलाऊँ,
तान, तान फण व्याल, कि तुझ पर मैं बाँसुरी बजाऊँ।

3

फूली सृष्टि नाद-बन्धन पर, अब तक फूल रही है,
बंसी के स्वर के धागे में धरती झूल रही है।

आदि छोर पर जो स्वर फूँका, पहुँचा अन्त तलक है,
तार-तार में गूँज गीत की, कण-कण बीच झलक है।
आलापों पर उठा जगत् को भर-भर पेंग झुलाऊँ,
तान, तान फण व्याल, कि तुझ पर मैं बाँसुरी बजाऊँ।

4

जगमग ओस-बिन्दु गुँथ जाते साँसों के तारों में,
गीत बदल जाते अनजाने मोती के हारों में।
जब-जब उठता नाद, मेघ मण्डलाकार घिरते हैं,
आस-पास बंसी के गीले इन्द्रधनुष तिरते हैं।
बाँधूँ मेघ कहाँ बंसी पर ? सुरधनु कहाँ सजाऊँ ?
तान, तान फण व्याल, कि तुझ पर मैं बाँसुरी बजाऊँ।

5

इस बंसी के मधुर नाद पर माया डोल चुकी है,
पटावरण कर दूर भेद अन्तर का खोल चुकी है।
झूम चुकी है प्रकृति चाँदनी में मादक गानों पर,
नचा चुका हूँ महानर्तकी को इस की तानों पर।
विषवर्षी पर अमृतवर्षिणी का जादू दिखलाऊँ,
तान, तान फण व्याल, कि तुझ पर मैं बाँसुरी बजाऊँ।

6

उड़े नाद के जो कण ऊपर, वे बन गये सितारे,
जो नीचे रह गये, कहीं हैं फूल, कहीं अंगारे।
भीगे अधर कभी वंशी के शीतल गंगाजल से;
कभी प्राण तक झुलस उठे हैं इस के हालाहल से।
शीतलता पीकर प्रहार से कैसे हृदय चुराऊँ ?
तान, तान फण व्याल, कि तुझ पर मैं बाँसुरी बजाऊँ।

7

यह बाँसुरी बजी मधु के सोते फूटे मधुवन में,
यह बाँसुरी बजी, हरियाली दौड़ गयी कानन में।

यह बाँसुरी बजी, प्रत्यागत हुए विहंग गगन से,
यह बाँसुरी बजी, सटकर विधु चलने लगा भुवन से,
अमृत-सरोवर में धो-धो तेरा भी जहर बहाऊँ,
तान, तान फण व्याल, कि तुझ पर मैं बाँसुरी बजाऊँ।

8

यह बाँसुरी बजी पनघट पर कालिन्दी के तट में,
यह बाँसुरी बजी मुरदों के आसन पर मरघट में।
बजी निशा के बीच आलुलायित केशों के तम में,
बजी सूर्य के साथ यही बाँसुरी रक्त-कर्दम में।
कालियदह में मिले हुए विष को पीयूष बनाऊँ,
तान, तान फण व्याल, कि तुझ पर मैं बाँसुरी बजाऊँ।

9

फूँक-फूँक विष-लपट, उगल जितना हो ज़हर हृदय में,
यह बंसी निर्गरल, बजेगी सदा क्षान्ति की लय में।
पहचाने किस तरह भला तू निज विष का मतवाला ?
मैं हूँ साँपों की पीठों पर कुसुम लादनेवाला।
विषदह से चल निकल, फूल से तेरा अंग सजाऊँ।
तान, तान फण व्याल, कि तुझ पर मैं बाँसुरी बजाऊँ।

10

ओ शंका के व्याल ! देख मत मेरे श्याम वदन को,
चक्षुःश्रवा ! श्रवण कर वंशी के भीतर के स्वन को।
जिस ने दिया तुझे विष, उस ने मुझ को गान दिया है,
ईर्ष्या तुझे उसी ने मुझ को भी अभिमान दिया है।
इस आशिष के लिए भाग्य पर क्यों न अधिक इतराऊँ ?
तान, तान फण व्याल, कि तुझ पर मैं बाँसुरी बजाऊँ।

11

विषधारी ! मत डोल कि मेरा आसन बहुत कड़ा है,
कृष्ण आज लघुता में भी साँपों से बहुत बड़ा है,
आया हूँ बाँसुरी बीच उद्धार लिये जनगण का,
फण पर तेरे खड़ा हुआ हूँ भार लिये त्रिभुवन का।
बढ़ा, बढ़ा नासिका, रन्ध्र में मुक्ति-सूत्र पहनाऊँ,
तान, तान फण व्याल, कि तुझ पर मैं बाँसुरी बजाऊँ।

कवि की मृत्यु

जब गीतकार मर गया, चाँद रोने आया,
चाँदनी मचलने लगी कफ़न बन जाने को।
मलयानिल ने शव को कन्धों पर उठा लिया,
वन ने भेजे चन्दन-श्रीखण्ड जलाने को।

सूरज बोला, यह बड़ी रोशनीवाला था,
मैं भी न जिसे भर सका कभी उजियाली से;
रँग दिया आदमी के भीतर की दुनिया को
इस गायक ने अपने गीतों की लाली से !

बोला बूढ़ा आकाश ध्यान जब यह धरता,
मुझ में यौवन का नया वेग जग जाता था।
इस के चिन्तन में डुबकी एक लगाते ही,
तन कौन कहे, मन भी मेरा रँग जाता था।

देवों ने कहा, बड़ा सुख था इसके मन की
गहराई में डूबने और उतराने में।
माया बोली, मैं कई बार थी भूल गयी
अपने को गोपन भेद इसे बतलाने में।

योगी था, बोला सत्य, भागता मैं फिरता,
यह जाल बढ़ाये हुए दौड़ता चलता था।

जब-जब लेता यह पकड़ और हँसने लगता,
धोखा देकर मैं अपना रूप बदलता था।

मर्दों को आयीं याद बाँकपन की बातें,
बोले, जो हो, आदमी बड़ा अलबेला था।
जिस के आगे तूफ़ान अदब से झुकते हैं,
उस को भी इसने अहंकार से झेला था।

नारियाँ बिलखने लगीं बाँसुरी के भीतर
जादू था, कोई अदा बड़ी मतवाली थी,
गर्जन में भी थी नमी, आग से भरे हुए
गीतों में भी कुछ चीज़ रुलानेवाली थी।

वे बड़ी-बड़ी आँखें आँसू से भरी हुईं,
पानी में जैसे कमल डूब उतराता हो।
वह मस्ती में झूमते हुए उस का आना,
मानो, अपना ही तनय झूमता आता हो।

चिन्तन में डूबा हुआ, सरल, भोला-भाला
बालक था, कोई पुरुष दिव्य अवतारी था।
तुम तो कहते हो मर्द, मगर, मन के भीतर
यह कलावन्त हम से भी बढ़कर नारी था।

चुपचाप ज़िन्दगी भर इस ने जो जुल्म सहे,
उतना नारी भी कहाँ मौन हो सहती है ?
आँखों के आँसू मन के भेद जता जाते,
कुछ सोच-समझ जिह्वा चाहे चुप रहती है।

पर, इसे नहीं रोने का भी अवकाश मिला,
सारा जीवन कट गया आग सुलगाने में।
आखिर, वह भी सो गया ज़िन्दगी ने जिस को,
था लगा रखा सोतों को छेड़ जगाने में।

बेबसी बड़ी उन बेचारों की क्या कहिए !
चुपचाप जिन्हें जीवन भर जलना होता है।
ऊपर-नीचे द्वेषों के कुन्त तने होते,
बचकर उन को बेदाग़ निकलना होता है।

जाओ, कवि, जाओ, मिला तुम्हें जो कुछ हम से,
दानी को उस के सिवा नहीं कुछ मिलता है।
चुन-चुनकर हम तोड़ते वही टहनी केवल
जिस पर कोई अपरूप कुसुम आ खिलता है।

विष के प्याले का मोल और क्या हो सकता ?
प्रेमी तो केवल मधुर प्रीत ही देता है।
कवि को चाहे संसार भेंट दे जो, लेकिन,
बदले में वह निष्कपट गीत ही देता है।

आवरण गिरा, जगती की सीमा शेष हुई,
अब पहुँच नहीं तुम तक इन हाहाकारों की।
नीचे की महफ़िल उजड़ गयी, ऊपर कल से
कुछ और चमक उट्ठेगी सभा सितारों की।

जनतन्त्र का जन्म

(26 जनवरी, 1950 ई.)

सदियों की ठण्डी-बुझी राख सुगबुगा उठी,
मिट्टी सोने का ताज पहन इठलाती है;
दो राह, समय के रथ का घर्घर-नाद सुनो,
सिंहासन ख़ाली करो कि जनता आती है।

जनता ? हाँ, मिट्टी की अबोध मूरतें वही,
जाड़े-पाले की कसक सदा सहनेवाली,
जब अंग-अंग में लगे साँप हों चूस रहे
तब भी न कभी मुँह खोल दर्द कहनेवाली।

जनता ? हाँ, लम्बी-बड़ी जीभ की वही कसम,
"जनता, सचमुच ही, बड़ी वेदना सहती है।"
"सो ठीक, मगर, आख़िर, इस पर जनमत क्या है?"
"है प्रश्न गूढ़; जनता इस पर क्या कहती है ?"

मानो, जनता हो फूल जिसे एहसास नहीं,
जब चाहो तभी उतार सजा लो दोनों में;
अथवा कोई दुधमुँही जिसे बहलाने के
जन्तर-मन्तर सीमित हों चार खिलौनों में।

लेकिन, होता भूडोल, बवण्डर उठते हैं,
जनता जब कोपाकुल हो भृकुटि चढ़ाती है;

दो राह, समय के रथ का घर्घर-नाद सुनो,
सिंहासन ख़ाली करो कि जनता आती है।

हुंकारों से महलों की नींव उखड़ जाती,
साँसों के बल से ताज हवा में उड़ता है,
जनता की रोके राह, समय में ताव कहाँ ?
वह जिधर चाहती, काल उधर ही मुड़ता है।

अब्दों, शताब्दियों, सहस्राब्द का अन्धकार
बीता; गवाक्ष अम्बर के दहके जाते हैं;
यह और नहीं कोई, जनता के स्वप्न अजय
चीरते तिमिर का वक्ष उमड़ते जाते हैं।

सब से विराट् जनतन्त्र जगत् का आ पहुँचा,
तैंतीस कोटि-हित सिंहासन तैयार करो;
अभिषेक आज राजा का नहीं, प्रजा का है,
तैंतीस कोटि जनता के सिर पर मुकुट धरो।

आरती लिये तू किसे ढूँढ़ता है मूरख,
मन्दिरों, राजप्रासादों में, तहखानों में ?
देवता कहीं सड़कों पर गिट्टी तोड़ रहे,
देवता मिलेंगे खेतों में, खलिहानों में।

फावड़े और हल राजदण्ड बनने को हैं,
धूसरता सोने से शृंगार सजाती है;
दो राह, समय के रथ का घर्घर-नाद सुनो,
सिंहासन ख़ाली करो कि जनता आती है।

शबनम की ज़ंजीर

रचना तो पूरी हुई, जान भी है इस में ?
पूछूँ जो कोई बात, मूर्ति बतलाएगी ?
लग जाय आग यदि किसी रोज़ देवालय में
चौंकेगी या यह खड़ी-खड़ी जल जाएगी ?

ढाँचे में तो सब ठीक-ठीक उतरा, लेकिन,
बेजान बुतों के कारीगर, कुछ होश करो;
जब तक पत्थर के भीतर साँस नहीं चलती,
सौगन्ध इसी की तुम्हें, न तुम सन्तोष करो।

भर सको अगर तो प्रतिमा में चेतना भरो,
यदि नहीं, निमन्त्रण दो जीवन के दानी को।
विभ्राट् महाबल जहाँ थके-से दीख रहे,
आगे आने दो वहाँ क्षीणबल प्राणी को।

तैरता हवा में जो, वह क्या भारी होगा ?
सपनों के तो सारथी क्षीणबल होते हैं;
संसार पुष्प से अपने को भूषित करता,
ये गन्धभार अपनी आत्मा में ढोते हैं।

सपनों का वह सारथी, यान जिसका कोमल,
आँखों से ओझल हृदय-हृदय में चलता है,

जिसके छूते ही मन की पलक उधर जाती,
विश्वास भ्रान्ति को भेद दीप-सा बलता है।

सपनों का वह सारथी, रात की छाया में,
आते जिस की श्रुति में संवाद सितारों से,
सरिताएँ जिस से अपना हाल कहा करतीं,
बातें करता जो फूलों और पहाड़ों से।

पपड़ियाँ तोड़ फूटते ज़िन्दगी के सोते,
रथ के चक्के की लीक जहाँ भी पड़ती है।
प्रतिमा सजीव होकर चलने-फिरने लगती,
मिट्टी की छाती में चेतना उमड़ती है।

छेनी-टाँकी क्या करें ? ज़िन्दगी की साँसें
लोहे पर धरकर नहीं बनाई जाती हैं;
धाराएँ जो मानव को उद्वेलित करतीं,
यन्त्रों के बल से नहीं बहाई जाती हैं।

विज्ञान काम कर चुका; हाथ उसका रोको;
आगे आने दो गुणी ! कला कल्याणी को।
जो भार नहीं विभ्राट्, महाबल उठा सके,
दो उसे उठाने किसी क्षीणबल प्राणी को।

मानव-मन को बेधते फूल के दल केवल,
आदमी नहीं कटता बरछों से, तीरों से;
लोहे की कड़ियों की साज़िश बेकार हुई,
बाँधो मनुष्य को शबनम की ज़ंजीरों से।

लोहे के पेड़ हरे होंगे

लोहे के पेड़ हरे होंगे, तू गान प्रेम का गाता चल,
नम होगी यह मिट्टी ज़रूर, आँसू के कण बरसाता चल।

सिसकियों और चीत्कारों से, जितना भी हो आकाश भरा,
कंकालों का हो ढेर, खप्परों से चाहे हो पटी धरा।
आशा के स्वर का भार, पवन को लेकिन, लेना ही होगा,
जीवित सपनों के लिए मार्ग मुर्दों को देना ही होगा।
रंगों के सातों घट उँड़ेल, यह अँधियारी रँग जाएगी,
ऊषा को सत्य बनाने को जावक नभ पर छितराता चल।

आदर्शों से आदर्श भिड़े, प्रज्ञा प्रज्ञा पर टूट रही।
प्रतिमा प्रतिमा से लड़ती है, धरती की किस्मत फूट रही।
आवर्तों का है विषम जाल, निरुपाय बुद्धि चकराती है,
विज्ञान-यान पर चढ़ी हुई सभ्यता डूबने जाती है।
जब-जब मस्तिष्क जयी होता, संसार ज्ञान से चलता है,
शीतलता की है राह हृदय, तू यह संवाद सुनाता चल।

सूरज है जग का बुझा-बुझा, चन्द्रमा मिलन-सा लगता है,
सब की कोशिश बेकार हुई, आलोक न इन का जगता है,
इन मलिन ग्रहों के प्राणों में कोई नवीन आभा भर दे,
जादूगर ! अपने दर्पण पर घिसकर इन को ताजा कर दे।
दीपक के जलते प्राण, दिवाली तभी सुहावन होती है,
रोशनी जगत् को देने को अपनी अस्थियाँ जलाता चल।

क्या उन्हें देख विस्मित होना, जो हैं अलमस्त बहारों में,
फूलों को जो हैं गूँथ रहे सोने-चाँदी के तारों में ?
मानवता का तू विप्र, गन्ध-छाया का आदि पुजारी है,
वेदना-पुत्र ! तू तो केवल जलने भर का अधिकारी है।
ले बड़ी खुशी से उठा, सरोवर में जो हँसता चाँद मिले,
दर्पण में रचकर फूल, मगर उस का भी मोल चुकाता चल।

काया की कितनी धूम-धाम ! दो रोज चमक बुझ जाती है;
छाया पीती पीयूष, मृत्यु के ऊपर ध्वजा उड़ाती है।
लेने दे जग को उसे, ताल पर जो कलहंस मचलता है,
तेरा मराल जल के दर्पण में नीचे-नीचे चलता है।
कनकाभ धूल झर जाएगी, ये रंग कभी उड़ जाएँगे,
सौरभ है केवल सार, उसे तू सब के लिए जुगाता चल।

क्या अपनी उन से होड़ अमरता की जिन को पहचान नहीं,
छाया से परिचय नहीं, गन्ध के जग का जिन को ज्ञान नहीं?
तो चतुर चाँद का रस निचोड़ प्यालों में ढाला करते हैं,
भट्ठियाँ चढ़ाकर फूलों से जो इत्र निकाला करते हैं।
ये भी जाएँगे कभी, मगर, आधी मनुष्यतावालों पर,
जैसे मुसकाता आया है, वैसे अब भी मुसकाता चल।

सभ्यता-अंग पर क्षत कराल, यह अर्थ-मानवों का बल है,
हम रोकर भरते उसे, हमारी आँखों में गंगाजल है।
शूली पर चढ़ा मसीहा को वे फूले नहीं समाते हैं,
हम शव को जीवित करने को छायापुर में ले जाते हैं।
भींगी चाँदनियों में जीता, जो कठिन धूप में मरता है,
उजियाली से पीड़ित नर के मन में गोधूलि बसाता चल।

यह देख नयी लीला उन की, फिर उन ने बड़ा कमाल किया,
गाँधी के लोहू से सारे, भारत-सागर को लाल किया।

जी उठे राम, जी उठे कृष्ण, भारत की मिट्टी रोती है,
क्या हुआ कि प्यारे गाँधी की यह लाश न जिन्दा होती है ?
तलवार मारती जिन्हें, बाँसुरी उन्हें नया जीवन देती,
जीवनी-शक्ति के अभिमानी ! यह भी कमाल दिखलाता चल।

धरती के भाग हरे होंगे, भारती अमृत बरसाएगी,
दिन की कराल दाहकता पर चाँदनी सुशीतल छाएगी।
ज्वालामुखियों के कण्ठों में कलकण्ठी का आसन होगा,
जलदों से लदा गगन होगा, फूलों से भरा भुवन होगा।
बेजान, यन्त्र-विरचित गूँगी, मूर्त्तियाँ एक दिन बोलेंगी,
मुँह खोल-खोल सब के भीतर शिल्पी ! तू जीभ बिठाता चल।

पुरूरवा की उक्ति

कौन है अंकुश, इसे मैं भी नहीं पहचानता हूँ।
पर, सरोवर के किनारे कण्ठ में जो जल रही है,
उस तृषा, उस वेदना को जानता हूँ।

आग है कोई, नहीं जो शान्त होती;
और खुलकर खेलने से भी निरन्तर भागती है।

रूप का रसमय निमन्त्रण
या कि मेरे ही रुधिर की वह्नि
मुझ को शान्ति से जीने न देती।
हर घड़ी कहती, उठो,
इस चन्द्रमा को हाथ से धरकर निचोड़ो,
पान कर लो यह सुधा, मैं शान्त हूँगी।
अब नहीं आगे कभी उद्भ्रान्त हूँगी।

किन्तु रस के पात्र पर ज्यों ही लगाता हूँ अधर को,
घूँट या दो घूँट पीते ही
न जाने, किस अतल से नाद यह आता,
"अभी तक भी न समझा ?
दृष्टि का जो पेय है, वह रक्त का भोजन नहीं है।
रूप की आराधना का मार्ग आलिंगन नहीं है।"

टूट गिरती हैं उमंगें,
बाहुओं का पाश हो जाता शिथिल है।

अप्रतिभ मैं फिर उसी दुर्गम जलधि में डूब जाता,
फिर वही उद्विग्न चिन्तन,
फिर वही पृच्छा चिरन्तन,
"रूप की आराधना का मार्ग
आलिंगन नहीं, तो और क्या है ?
स्नेह का सौन्दर्य को उपहार
रस-चुम्बन नहीं, तो और क्या है ?"

रक्त की उत्तप्त लहरों की परिधि के पार
कोई सत्य हो तो,
चाहता हूँ, भेद उसका जान लूँ।
पन्थ हो सौन्दर्य की आराधना का व्योम में यदि
शून्य की उस रेख को पहचान लूँ।

पर जहाँ तक भी उड़ूँ, इस प्रश्न का उत्तर नहीं है।
मृत्ति महदाकाश में ठहरे कहाँ पर ? शून्य है सब।
और नीचे भी नहीं सन्तोष,
मिट्टी के हृदय से
दूर होता ही कभी अम्बर नहीं है।

इस व्यथा को झेलता
आकाश की निस्सीमता में
घूमता फिरता विकल, विभ्रान्त,
पर कुछ भी न पाता।
प्रश्न जो कढ़ता,
गगन की शून्यता में गूँजकर सब ओर
मेरे ही श्रवण में लौट आता।

और इतने में मही का गान फिर देता सुनाई,
''हम वही जग हैं, जहाँ पर फूल खिलते हैं।
दूब है शैया हमारे देवता की,
पुष्प के वे कुंज मन्दिर हैं,
जहाँ शीतल, हरित, एकान्त मण्डप में प्रकृति के
कण्टकित युवती-युवक स्वच्छन्द मिलते हैं।

''इन कपोलों की ललाई देखते हो ?
और अधरों की हँसी यह कुन्द-सी, जूही-कली-सी ?
गौर चम्पक-यष्टि-सी यह देह श्लथ पुष्पाभरण से,
स्वर्ण की प्रतिमा कला के स्वप्न-साँचे में ढली-सी ?

''यह तुम्हारी कल्पना है, प्यार कर लो।
रूपसी नारी प्रकृति का चित्र है सब से मनोहर।
और गगनचारी ! यहाँ मधुमास छाया है।
भूमि पर उतरो,
कमल, कर्पूर, कुंकुम से, कुटज से
इस अतुल सौन्दर्य का श्रृंगार कर लो।''

गीत आता है मही से—
या कि मेरे ही रुधिर का राग ?
यह उठता गगन में ?
बुलबुलों-सी फूटने लगतीं मधुर स्मृतियाँ हृदय में।
याद आता है मदिर उल्लास में फूला हुआ वन,
याद आते हैं तरंगित अंग के रोमांच, कम्पन;
स्वर्णवर्णा वल्लरी में फूल-से खिलते हुए सुख,
याद आता है निशा के ज्वार में उन्माद का मुख।
कामनाएँ प्राण को हिलकोरती हैं।
चुम्बनों के चिह्न जग पड़ते त्वचा में।

फिर किसी का स्पर्श पाने को तृषा चीत्कार करती

मैं न रुक पाता कहीं,
फिर लौट आता हूँ पिपासित
शून्य से साकार सुषमा के भुवन में,
युद्ध से भागे हुए उस् वेदना-विह्वल युवक-सा
जो कहीं रुकता नहीं,
बेचैन जा गिरता अकुण्ठित
तीर-सा सीधे प्रिया की गोद में।

चूमता हूँ दूब को, जल को, प्रसूनों, पल्लवों को,
वल्लरी को बाँह भर उर से लगाता हूँ;
बालकों-सा मैं तुम्हारे वक्ष में मुँह को छिपाकर
नींद की निस्तब्धता में डूब जाता हूँ।

नींद जल का स्रोत है, छाया सघन है,
नींद श्यामल मेघ है, शीतल पवन है।

किन्तु जगकर देखता हूँ,
कामनाएँ वर्तिका-सी बल रही हैं,
जिस तरह पहले पिपासा से विकल थीं,
प्यास से आकुल अभी भी जल रही हैं।

रात भर, मानो, उन्हें दीपक-सदृश जलना पड़ा हो,
नींद में, मानो, किसी मरुदेश में चलना पड़ा हो।

फिर क्षुधित कोई अतिथि आवाज देता,
फिर अधर-पुट खोजने लगते अधर को,
कामना छूकर त्वचा को फिर जगाती है।
रेंगने लगते सहस्रों साँप सोने के रुधिर में,
चेतना रस की लहर में डूब जाती है।

और तब सहसा

न जाने, ध्यान खो जाता कहाँ पर।
सत्य ही, रहता नहीं यह ज्ञान,
तुम कविता, कुसुम या कामिनी हो।
आरती की ज्योति को भुज में समेटे
मैं तुम्हारी ओर अपलक देखता एकान्त मन में
रूप के उद्‌गम अगम का भेद गुनता हूँ।

साँस में सौरभ, तुम्हारे वर्ण में गायन भरा है।
सींचता हूँ प्राण को इस गन्ध की भीनी लहर से,
और अंगों की विभा की वीचियों से एक होकर
मैं तुम्हारे रंग का संगीत सुनता हूँ।

और फिर यह सोचने लगता, कहाँ, किस लोक में हूँ।
कौन है यह वन सघन हरियालियों का,
झूमते फूलों, लचकती डालियों का ?
कौन है यह देश, जिस की स्वामिनी मुझ को निरन्तर
वारुणी की धार से नहला रही है ?
कौन है यह जग, समेटे अंक में ज्वालामुखी को
चाँदनी चुमकार कर बहला रही है ?

कौमुदी के इस सुनहरे जाल का बल तोलता हूँ,
एक पल उड्डीन होने के लिए पर खोलता हूँ।

पर, प्रभंजन मत्त है इस भाँति रस-आमोद में,
उड़ न सकता, लौट गिरता है कुसुम की गोद में।

टूटता तोड़े नहीं यह किसलयों का दाम,
फूलों की लड़ी जो बँध गयी, खुलती नहीं है !

कामनाओं के झकोरे रोकते हैं राह मेरी,
खींच लेती है तृषा पीछे पकड़कर बाँह मेरी।

सिन्धु-सा उद्दाम, अपरम्पार मेरा बल कहाँ है ?
गूँजता जिस शक्ति का सर्वत्र जयजयकार,
उस अटल संकल्प का सम्बल कहाँ है ?

यह शिला-सा वक्ष, ये चट्टान-सी मेरी भुजाएँ,
सूर्य के आलोक से दीपित, समुन्नत भाल,
मेरे प्राण का सागर अगम, उत्ताल, उच्छल है।

सामने टिकते नहीं वनराज, पर्वत डोलते हैं,
काँपता है कुण्डली मारे समय का व्याल,
मेरी बाँह में मारुत, गरुड़, गजराज का बल है।

मर्त्य मानव की विजय का तूर्य हूँ मैं,
उर्वशी ! अपने समय का सूर्य हूँ मैं।
अन्ध तम के भाल पर पावक जलाता हूँ,
बादलों के सीस पर स्यन्दन चलाता हूँ।

पर, न जाने, बात क्या है !
इन्द्र का आयुध पुरुष जो झेल सकता है,
सिंह से बाँहें मिलाकर खेल सकता है,
फूल के आगे वही असहाय हो जाता,
शक्ति के रहते हुए निरुपाय हो जाता।

विद्ध हो जाता सहज बंकिम नयन के बाण से,
जीत लेती रूपसी नारी उसे मुसकान से।

मैं तुम्हारे बाण का बींधा हुआ खग,
वक्ष पर धर सीस मरना चाहता हूँ।

मैं तुम्हारे हाथ का लीला-कमल हूँ,
प्राण के सर में उतरना चाहता हूँ।

कौन कहता है,
तुम्हें मैं छोड़कर आकाश में विचरण करूँगा ?

बाहुओं के इस वलय में गात्र ही बन्दी नहीं है।
वक्ष के इस तल्प पर सोती न केवल देह,
मेरे व्यग्र, व्याकुल प्राण भी विश्राम पाते हैं।

मर्त्य नर को देवता कहना मृषा है।
देवता शीतल, मनुज अंगार है।

देवताओं की नदी में ताप की लहरें न उठतीं,
किन्तु नर के रक्त में ज्वालामुखी हुंकारता है,
घूर्णियाँ चिनगारियों की नाचती हैं,
नाचते उड़कर दहन के खण्ड पत्तों-से हवा में,
मानवों का मन गले-पिघले अनल की धार है।

चाहिए देवत्व,
पर, इस आग को धर दूँ कहाँ पर ?
कामनाओं को विसर्जित व्योम में कर दूँ कहाँ पर ?
वह्नि का बेचैन यह रसकोष, बोलो, कौन लेगा ?
आग के बदले मुझे सन्तोष, बोलो, कौन देगा ?

फिर दिशाएँ मौन, फिर उत्तर नहीं है।

प्राण की चिर-संगिनी यह वह्नि,
इस को साथ लेकर
भूमि से आकाश तक चलते रहो।
मर्त्य नर का भाग्य !

जब तक प्रेम की धारा न मिलती,
आप अपनी आग में जलते रहो।

एक ही आशा, मरुस्थल की तपन में
ओ सजल कादम्बिनी ! सिर पर तुम्हारी छाँह है।
एक ही सुख है, उरस्थल से लगा हूँ,
ग्रीव के नीचे तुम्हारी बाँह है।

इन प्रफुल्लित प्राण-पुष्पों में मुझे शाश्वत शरण दो।
गन्ध के इस लोक से बाहर न जाना चाहता हूँ।
मैं तुम्हारे रक्त के कण में समाकर
प्रार्थना के गीत गाना चाहता हूँ।

उर्वशी की उक्ति

पर, क्या बोलूँ ? क्या कहूँ ?
भ्रान्ति यह देह-भाव।
मैं मनोदेश की वायु व्यग्र, व्याकुल, चंचल;
अवचेत प्राण की प्रभा, चेतना के जल में
मैं रूप-रंग-रस-गन्ध-पूर्ण साकार कमल।

मैं नहीं सिन्धु की सुता;
तलातल-अतल-वितल-पाताल छोड़,
नीले समुद्र को फोड़ शुभ्र, झलमल फेनांशुक में प्रदीप्त
नाचती ऊर्मियों के सिर पर
मैं नहीं महातल से निकली।

मैं नहीं गगन की लता
तारकों में पुलकित फूलती हुई,
मैं नहीं व्योमपुर की बाला,
विधु की तनया, चन्द्रिका-संग,
पूर्णिमा-सिन्धु की परमोज्ज्वल आभा-तरंग,
मैं नहीं किरण के तारों पर झूलती हुई भू पर उतरी।

मैं नाम-गोत्र से रहित पुष्प,
अम्बर में उड़ती हुई मुक्त आनन्द-शिखा
इतिवृत्तिहीन,

सौन्दर्य-चेतना की तरंग;
सुर-नर-किन्नर-गन्धर्व नहीं,
प्रिय ! मैं केवल अप्सरा
विश्वनर के अतृप्त इच्छा-सागर से समुद्भूत।

कामना तरंगों से अधीर
जब विश्वपुरुष का हृदय-सिन्धु
आलोडित, क्षुभित, मथित होकर,
अपनी समस्त बड़वाग्नि
कण्ठ में भरकर मुझे बुलाता है,
तब मैं अपूर्वयौवना
पुरुष के निभृत प्राणतल से उठकर
प्रसरित करती निर्वसन, शुभ्र, हेमाभ कान्ति
कल्पना-लोक से उतर भूमि पर आती हूँ,
विजयिनी विश्वनर को अपने उत्तुंग वक्ष
पर सुला अमित कल्पों के अश्रु सुखाती हूँ।

जन-जन के मन की मधुर वह्नि, प्रत्येक हृदय की उजियाली,
नारी की मैं कल्पना चरम नर के मन में बसनेवाली।

विषधर के फण पर अमृतवर्ति;
उद्धत, अदम्य, बर्बर बल पर
रूपांकुश, क्षीण मृणाल-तार।
मेरे सम्मुख नत हो रहते गजराज मत्त;
केसरी, शरभ, शार्दूल भूल निज हिंस्र भाव
गृह-मृग-समान निर्विष, अहिंस्र बनकर जीते।
मेरी भ्रू-स्मिति को देख चकित, विस्मित, विभोर
शूरमा निमिष खोले अवाक् रह जाते हैं;
श्लथ हो जाता स्वयमेव शिंजिनी का कसाव,
संत्रस्त करों से धनुष-बाण गिर जाते हैं।

कामना-वह्नि की शिखा मुक्त मैं अनवरुद्ध,
मैं अप्रतिहत, मैं दुर्निवार;
मैं सदा घूमती फिरती हूँ

पवनान्दोलित वारिद-तरंग पर समासीन
नीहार-आवरण में अम्बर के आर-पार;
उड़ते मेघों को दौड़ बाहुओं में भरती,
स्वप्नों की प्रतिमाओं का आलिंगन करती।

विस्तीर्ण सिन्धु के बीच शून्य, एकान्त द्वीप,
यह मेरा उर।

देवालय में देवता नहीं, केवल मैं हूँ।
मेरी प्रतिमा को घेर उठ रही अगुरु-गन्ध,
बज रहा अर्चना में मेरी मेरा नूपुर।

मैं कला-चेतना का मधुमय, प्रच्छन्न स्रोत,
रेखाओं में अंकित कर अंगों के उभार,
भंगिमा, तरंगित वर्तुलता, वीचियाँ लहर,
तन की प्रकान्ति रंगों में लिये उतरती हूँ।

पाषाणों के अनगढ़ अंगों को काट-छाँट

मैं ही निविडस्तननता, मुष्टिमध्यमा,
मदिरलोचना, कामलुलिता नारी
प्रस्तरावरण कर भंग,
तोड़ तम को उन्मत्त उभरती हूँ।

भू-नभ का सब संगीत नाद मेरे निस्सीम प्रणय का है,
सारी कविता जयगान एक मेरी त्रयलोक-विजय का है।
प्रिय मुझे प्रखर कामना-कलित, सन्तप्त, व्यग्र, चंचल चुम्बन,

प्रिय मुझे रसोदधि में निमग्न उच्छल, हिल्लोल-निरत जीवन।
तारों की झिलमिल छाया में फूलों की नाव बहाती हूँ,
मैं नैश प्रभा, सब के भीतर निश की कल्पना जगाती हूँ।

मादन सुगन्ध पवमान-दलित सन्ध्या-तन से उठनेवाली
नभ से आलिंगित कुमुदवती चन्द्रिका-यामिनी मतवाली,

कबरी के फूलों का सुवास, आकुंचित अधरों का कम्पन,
परिरम्भ वेदना से विभोर, कण्टकित अंग, मधुमत्त नयन;

दो प्राणों से उठनेवाली वे झंकृतियाँ गोपन, मधुमय,
जो अगरु-धूम-सी हो जातीं ऊपर उठ एक अपर में लय;

दो दीपों की सम्मिलित ज्योति, वह एक शिखा जब जगती है।
मन के अगाध रत्नाकर में यह देह डूबने लगती है।

दो हृदयों का वह मूक मिलन, तन शिथिल, स्रस्त अतिशय सुख से।
अलसित आँखें देखतीं, न कोई शब्द निकलता है मुख से।

कितनी पावन वह रस-समाधि ! जब सेज स्वर्ग बन जाती है,
गोचर शरीर में विभा अगोचर सुख की झलक दिखाती है !

देवता एक है शयित कहीं इस मदिर शान्ति की छाया में,
आरोहण के सोपान लगे हैं त्वचा, रुधिर में काया में।
परिरम्भ-पाश में बँधे हुए उस अम्बर तक उठ जाओ रे !
देवता मेम का सोया है, चुम्बन से उसे जगाओ रे !

चिन्तन की लहरों के समान सौन्दर्य-लहर में भी है बल,
सातों अम्बर तक उड़ता है रूपसी नारि का स्वर्णांचल।
जिस मधुर भूमिका में जन को दर्शन-तरंग पहुँचाती है,
उस दिव्य लोक तक हमें प्रेम की नाव सहज ले जाती है।

ओ शून्य पवन में मुझे देख चुम्बन अर्पित करनेवालो !
सम्पूर्ण निशा मेरी छवि का उन्निद्र ध्यान धरनेवालो !

मैं देश-काल से परे चिरन्तन नारी हूँ।
मैं आत्मतन्त्र यौवन की नित्य नवीन प्रभा,
रूपसी अमर मैं चिर-युवती सुकुमारी हूँ।
तुम त्रिभुवन में अथवा त्रिकाल में जहाँ कहीं भी हो,
अन्तर में धैर्य धरो।
सरिता, समुद्र, गिरि, वन मेरे व्यवधान नहीं।
मैं भूत, भविष्यत्, वर्तमान की कृत्रिम बाधा से विमुक्त;
मैं विश्वप्रिया।

तुम पन्थ जोहते रहो,
अचानक किसी रात मैं आऊँगी।
अधरों में अपनी अधरों की मदिरा उँड़ेल,
मैं तुम्हें वक्ष से लगा
युगों की संचित तपन मिटाऊँगी।

चेतना का शृंग

उर्वशी

अतल, अनादि, अनन्त, पूर्ण, बृंहित, अपार अम्बर में
सीमा खींचें कहाँ ? निमिष, पल, दिवस, मास, संवत्सर
महाकाश में टँगे काल के लक्तक-से लगते हैं।
प्रिय ! उस पत्रक को समेट लो, जिस में समय सनातन
क्षण, मुहूर्त, संवत् शताब्दि की बूँदों में अंकित है।
बहने दो निश्चेत शान्ति की इस अकूल धारा में,
देश-काल से परे, छूटकर अपने भी हाथों से।
किस समाधि का शिखर, चेतना जिस पर ठहर गयी है ?
उड़ता हुआ विशिख अम्बर में स्थिर-समान लगता है।

पुरूरवा

रुको समय-सरिते! पल! अनुपल! काल-शकल! घटिकाओ!
इस प्रकार, आतुर उड़ान भर कहाँ तुम्हें जाना है ?
कहीं समापन नहीं ऊर्ध्व-गामी जीवन की गति का,
काल-पयोनिधि का त्रिकाल में कोई कूल नहीं है।
कहीं कुण्डली मार बैठ जाओ नक्षत्र-निलय में,
मत ले जाओ खींच निशा को आज सूर्य-वेदी पर।

रुको, पान करने दो शीतलता शतपत्र कमल की;
एक सघन क्षण में समेटने दो विस्तार समय का,
एक पुरुष में भर त्रिकाल की सुरभि सूँघ लेने दो।

मिटा कौन ? जो बीत गया, पीछे की ओर खड़ा है,
जनमा अब तक नहीं, अभी वह घन के अँधियाले में
बैठा है सामने छन्न, पर, सब कुछ देख रहा है।

जिस प्रकार, नगराज जानता व्यथा विन्ध्य के डर को,
और हिमालय की गाथा है विदित महासागर को,
वर्तमान, त्यों ही, अपने गृह में जो कुछ करता है,
भूत और भवितव्य, उभय उस रचना के साखी हैं।

सिन्धु, विन्ध्य, हिमवान् खड़े हैं दिगायाम में जैसे
एक साथ; त्यों काल-देवता के महान् प्रांगण में
भूत, भविष्यत् वर्तमान, सब साथ-साथ ठहरे हैं
बातें करते हुए परस्पर गिरा-मुक्त भाषा में।

कहाँ देश, हम नहीं व्योम में जिसके गूँज रहे हैं ?
कौन कल्प, हम नहीं तैरते हैं जिस के सागर में ?
महाशून्य का उत्स हमारे मन का भी उद्गम है,
बहती है चेतना काल के आदि-मूल को छूकर।

उर्वशी

हम त्रिलोकवासी, त्रिकालचर, एकाकार समय से,
भूत, भविष्यत्, वर्तमान, तीनों के एकार्णव में,
तैर रहे सम्पृक्त सभी वीचियों, कणों, अणुओं से।
समा रहीं धड़कनें उरों की अप्रतिहत त्रिभुवन में,
काल-रन्ध्र भर रहा हमारी साँसों के सौरभ से।
अन्तर्नभ का यह प्रसार ! यह परिधि-भंग प्राणों का !
सुख की इस अपार महिमा को कहाँ समेट धरें हम ?

पुरूरवा

महाशून्य के अन्तर-गृह में, उस अद्वैत-भवन में,
जहाँ पहुँच दिक्एक काल हैं, कोई भेद नहीं है।

इस निरभ्र नीलान्तरिक्ष की निर्जर मंजूषा में,
सर्ग-प्रलय के पुरावृत्त जिस में समग्र संचित हैं।

दूरागत इस सतत-संचरण-मय समीर के कर में,
कथा आदि की जिसे अन्त की श्रुति तक ले जाना है।

इस प्रदीप्त निश के अंचल में जो आप्रलय निरन्तर,
इसी भाँति, सुनती जाएगी कूजन गूढ़ प्रणय का।

उस अदृश्य के पद पर, जिसकी दयासिक्त, मृदु स्मृति में
हम दोनों घूमते और क्रीड़ा-विहार करते हैं।

जिस की इच्छा का प्रसार भूतल, पाताल, गगन है,
दौड़ रहे नभ में अनन्त कन्दुक जिस की लीला के
अगणित सविता-सोम, अपरिमित ग्रह, उडु-मण्डल बनकर;
नारी बन जो स्वयं पुरुष को उद्वेलित करता है
और बेधता पुरुष-कान्ति बन हृदय-पुष्प नारी का।

निधि में जल, वन में हरीतिमा जिस का घनावरण है,
रक्त-मांस-विग्रह भंगुर ये उसी विभा के पट हैं।

प्रणय-शृंग की निश्चेतनता में अधीर बाँहों के
आलिंगन में देह नहीं श्लथ, यही विभा बँधती है।
और चूमते हम अचेत हो जब असंज्ञ अधरों को,
वह चुम्बन अदृश्य के चरणों पर भी चढ़ जाता है।

देह मृत्ति, दैहिक प्रकाश की किरणें मृत्ति नहीं हैं,
अधर नष्ट होते, मिटती झंकार नहीं चुम्बन की;
यह अरूप आभा-तरंग अर्पित उस के चरणों पर,
निराकार जो जाग रहा है सारे आकारों में।

उर्वशी

रोम-रोम में वृक्ष, तरंगित, फेनिल हरियाली पर
चढ़ी हुई आकाश-ओर मैं कहाँ उड़ी जाती हूँ ?

पुरूरवा

देह डूबने चली अतल मन के अकूल सागर में,
किरणें फेंक अरूप रूप को ऊपर खींच रहा है।

उर्वशी

करते नहीं स्पर्श क्यों पगतल मृत्ति और प्रस्तर का ?
सघन, उष्ण वह वायु कहाँ है ? हम इस समय कहाँ हैं ?

पुरूरवा

छूट गयी धरती नीचे, आभा की झंकारों पर
चढ़े हुए हम देह छोड़कर मन में पहुँच रहे हैं।

उर्वशी

फूलों का सम्पूर्ण भुवन सिर पर, इस तरह, उठाये
यह पर्वत का शृंग मुदित हम को क्यों हेर रहा है ?

पुरूरवा

अयुत युगों से ये प्रसून यों ही खिलते आये हैं,
नित्य जोहते पन्थ हमारे इसी महान् मिलन का।

जब भी तन की परिधि पार कर मन के उच्च निलय में,
नर-नारी मिलते समाधि-सुख के निश्चेत शिखर पर,
तब प्रहर्ष की अति से यों ही प्रकृति काँप उठती है,
और फूल यों ही प्रसन्न होकर हँसने लगते हैं

उर्वशी

जला जा रहा अर्थ सत्य का सपनों की ज्वाला में,
निराकार में आकारों की पृथ्वी डूब रही है।
यह कैसी माधुरी ? कौन स्वर लय में गूँज रहा है
त्वचा-जाल पर, रक्त-शिराओं में, अकूल अन्तर में ?
ये ऊर्मियाँ ! अशब्द नाद ! उफ् री, बेबसी गिरा की !
दोगे कोई शब्द ? कहूँ क्या कहकर इस महिमा को ?

पुरूरवा

शब्द नहीं हैं; यह गूँगे का स्वाद, अगोचर सुख है;
प्रणय-प्रज्वलित उर में जितनी झंकृतियाँ उठती हैं,
कहकर भी उन को कह पाते कहाँ सिद्ध प्रेमी भी ?
भाषा रूपाश्रित, अरूप है यह तरंग प्राणों की।

उर्वशी

कौन पुरुष तुम ?

पुरूरवा

जो अनेक कल्पों के अँधियाले में
तुम्हें खोजता फिरा तैरकर बारम्बार मरण को,
जन्मों के अनेक कुंजों, वीथियों, प्रार्थनाओं में,
पर तुम मिलीं एक दिन सहसा जिसे शुभ्र मेघों पर
एक पुष्प में अमित युगों के स्वप्नों की आभा-सी।

उर्वशी

और कौन मैं ?

पुरूरवा

ठीक-ठीक यह नहीं बता सकता हूँ।
इतना ही है ज्ञात, तुम्हारे आते ही अन्तर का

द्वार स्वयं खुल गया और प्राणों का निभृत निकेतन
अकस्मात् भर गया स्वरित रंगों के कोलाहल से।
जब से तुम आयीं, पृथ्वी कुछ अधिक मुदित लगती है;
शैल समझते हैं, उन के प्राणों में जो धारा है,
बहती है पहले से वह कुछ अधिक रसवती होकर।
जब से तुम आयीं, धरती पर फूल अधिक खिलते हैं,
दौड़ रही कुछ नयी दीप्ति-सी शीतल हरियाली में
सब हैं सुखी, एक नक्षत्रों को ऐसा लगता है,
जैसे कोई वस्तु हाथ से उन के निकल गयी हो।

परशुराम के उपदेश

1

सिखलाएगा वह, ऋत एक ही अनल हैं,
ज़िन्दगी नहीं वह, जहाँ नहीं हलचल है।
जिन में दाहकता नहीं, न तो गर्जन है,
सुख की तरंग का जहाँ अन्ध वर्जन है,
जो सत्य राख में सने, रूक्ष, रूठे हैं
छोड़ो उन को, वे सही नहीं, झूठे हैं।

2

वैराग्य छोड़ बाँहों की विभा सँभालो,
चट्टानों की छाती से दूध निकालो।
है रुकी जहाँ भी धार, शिलाएँ तोड़ो,
पीयूष चन्द्रमाओं को पकड़ निचोड़ो।
चढ़ तुंग शैल-शिखरों पर सोम पियो रे!
योगियों नहीं, विजयी के सदृश जियो रे!

3

जब कुपित काल धीरता त्याग जलता है,
चिनगी बन फूलों का पराग जलता है,
सौन्दर्य-बोध बन नयी आग जलता है।
ऊँचा उठकर कामार्त राग जलता है।

अम्बर पर अपनी विभा प्रबुद्ध करो रे !
गरजे कृशानु, तब कंचन शुद्ध करो रे !

4

जिस की बाँहें बलमयी, ललाट अरुण है,
भामिनी वही तरुणी, नर वही तरुण है।
है वही प्रेम जिस की तरंग उच्छल है,
वारुणी-धार में मिश्रित जहाँ गरल है।
उद्दाम प्रीति बलिदान-बीज बोती है;
तलवार प्रेम से और तेज होती है।

5

पी जिसे उमड़ता अनल, भुजा भरती है,
वह शक्ति सूर्य की किरणों में झरती है।
मरु के प्रदाह में छिपा हुआ जो रस है,
तूफ़ान-अन्धड़ों में जो अमृत-कलस है,
उस तपन-तत्त्व से हृदय-प्राण खींचो रे !
खींचो, भीतर आँधियाँ और खींचो रे !

6

छोड़ो मत अपनी आन, सीस कट जाए,
मत झुको अनय पर, भले व्योम फट जाए।
दो बार नहीं यमराज कण्ठ धरता है,
मरता है जो, एक ही बार मरता है।
तुम स्वयं मरण के मुख पर चरण धरो रे!
जीना हो मरने से नहीं डरो रे !

7

स्वातन्त्र्य जाति की लगन, व्यक्ति की धुन है,
बाहरी वस्तु यह नहीं, भीतरी गुण है।
नत हुए बिना जो अशनि-घात सहती है,
स्वाधीन जगत् में वही जाति रहती है।

वीरत्व छोड़ पर का मत चरण गहो रे !
जो पड़े आन, खुद ही सब आग सहो रे!

8

आँधियाँ नहीं जिस में उमंग भरती हैं,
छातियाँ जहाँ संगीनों से डरती हैं,
शोणित के बदले जहाँ अश्रु बहता है,
वह देश कभी स्वाधीन नहीं रहता है।
पकड़ो अयाल, अन्धड़ पर उछल चढ़ो रे!
किरिचों पर अपने तन का चाम मढ़ो रे!

9

हैं खड़े हिंस्र वृक-व्याघ्र, खड़ा पशुबल है,
ऊँची मनुष्यता का पथ नहीं सरल है।
ये हिंस्र साधु पर भी न तरस खाते हैं।
कण्ठी-माला के सहित चबा जाते हैं।
जो वीर काटकर इन्हें पार जाएगा,
उत्तुंग श्रृंग पर वही पहुँच पाएगा।

10

जो पुरुष भूल शायक, कुठार को, असि को,
पूजता मात्र चिन्तन, विचार को, मसि को,
सत्य का नहीं बहुमान किया करता है,
केवल सपनों का ध्यान किया करता है,
बस में उस के यह लोक न रह जाएगा।
है हवा स्वप्न, कर में वह क्यों आएगा ?

11

उपशम को ही जो जाति धर्म कहती है,
शम, दम, विराग को श्रेष्ठ कर्म कहती है,
धृति को प्रहार, क्षान्ति को वर्म कहती है,
अक्रोध, विनय को विजय-मर्म कहती है।

अपमान कौन, वह जिस को नहीं सहेगी?
सब को असीस, सब का बन दास रहेगी।

12

यह कठिन शाप सुकुमार धर्म-साधन का,
रण-विमुख, शान्त जीवन के आराधन का;
जातियाँ पावकों से बचकर चलती हैं,
निर्वीर्य कल्पनाएँ रचकर चलती हैं।
वृन्तों पर जलते सूर्य छोड़ देती हैं,
चुन-चुनकर केवल चाँद तोड़ लेती हैं।

13

दो उन्हें राम, तो मात्र नाम वे लेंगी,
विक्रमी शरासन से न काम वे लेंगी;
नवनीत बना देतीं भट अवतारी को,
मोहन मुरलीधर पांचजन्य-धारी को।
पावक को बुझा तुषार बना देती हैं,
गाँधी को शीतल क्षार बना देती हैं।

14

है सही, बना पहले पृथ्वी से जल था,
पर, बहुत पूर्व उस से बन चुका अनल था।
जब प्रथम-प्रथम हो उठा तत्त्व चंचल था,
प्रेरणा-स्रोत पर विनय नहीं थी, बल था।
है अनल ब्रह्म, पावक-तरंग जीवन है,
अब समझा, क्यों ज्वाला अभंग जीवन है?

15

स्वर में पावक यदि नहीं, वृथा वन्दन है,
वीरता नहीं, तो सभी विनय क्रन्दन है।

सिर पर जिस के असिघात, रक्त-चन्दन है,
भ्रामरी उसी का करती अभिनन्दन है।
दानवी रक्त से सभी पाप धुलते हैं,
ऊँची मनुष्यता के पथ भी खुलते हैं।

16

हैं दुखी मेष, क्यों लहू शेर चखते हैं,
नाहक इतने क्यों दाँत तेज़ रखते हैं।
पर, शेर द्रवित हो दशन तोड़ क्यों लेंगे ?
मेषों के हित व्याघ्रता छोड़ क्यों देंगे ?
एक ही पन्थ, तुम भी आघात हनो रे !
मेषत्व छोड़ मेषो ! तुम व्याघ्र बनो रे !

17

जो अड़े, शेर उस नर से डर जाता है,
है विदित, व्याघ्र को व्याघ्र नहीं खाता है।
सच पूछो तो अब भी सच यही वचन है,
सभ्यता क्षीण, बलवान् हिंस्र कानन है।
एक ही पन्थ अब भी जग में जीने का,
अभ्यास करो छागियो ! रक्त पीने का।

18

जब शान्तिवादियों ने कपोत छोड़े थे,
किस ने आशा से नहीं हाथ जोड़े थे ?
पर, हाय, धर्म यह भी धोखा है, छल है;
उजले कबूतरों में भी छिपा अनल है।
पंजों में इन के धार धरी होती है।
कइयों में तो बारूद भरी होती है।

19

हैं जिन्हें दाँत, उन से अदन्त कहते हैं,
यानी शूरों को देख सन्त कहते हैं,

'तुम तुड़ा दाँत क्यों पुण्य नहीं पाते हो ?'
यानी तुम भी क्यों भेड़ न बन जाते हो।
पर कौन शेर भेड़ों की बात सुनेगा,
ज़िन्दगी छोड़ मरने की राह चुनेगा ?

20

वे देश शान्ति के सब से शत्रु प्रबल हैं,
जो बहुत बड़े होने पर भी दुर्बल हैं,
हैं जिन के उदर विशाल, बाँह छोटी हैं,
भोथरे दाँत, पर, जीभ बहुत मोटी है।
औरों के पाले जो अलज्ज पलते हैं,
अथवा शेरों पर लदे हुए चलते हैं।

21

सिंहों पर अपना अतुल भार मत डालो,
हाथियो ! स्वयं अपना तुम बोझ सँभालो।
यदि लदे फिरे, यों ही, तो पछताओगे,
शव मात्र आप अपना तुम रह जाओगे।
यह नहीं मात्र अपकीर्ति, अनय की अति है।
जाने, कैसे सहती यह दृश्य प्रकृति है ?

22

उद्देश्य जन्म का नहीं कीर्ति या धन है,
सुख नहीं, धर्म भी नहीं, न तो दर्शन है,
विज्ञान, ज्ञान-बल नहीं, न तो चिन्तन है,
जीवन का अन्तिम ध्येय स्वयं जीवन है।
सब से स्वतन्त्र यह रस जो अनघ पिएगा,
पूरा जीवन केवल वह वीर जिएगा।

आपद्धर्म

अरे उर्वशीकार !
कविता की गरदन पर धरकर पाँव खड़ा हो।
हमें चाहिए गर्म गीत उन्माद, प्रलय का,
अपनी ऊँचाई से तू कुछ और बड़ा हो।

कच्चा पानी ठीक नहीं,
ज्वर-ग्रसित देश है !
उबला हुआ समुष्ण सलिल है पथ्य,
वही परिशोधित जल दे।
जाड़े की है रात, गीत की गरमाहट दे,
तप्त अनल दे।

रोज पत्र आते हैं, जलते गान लिखूँ मैं,
जितना हूँ, उस से कुछ अधिक जवान दिखूँ मैं।

और, सत्य ही, मैं भी युग के स्वरावेग से चूर,
दूर उर्वशी लोक से,
गयी जवानी की बुझती भट्ठी फिर सुलगाता हूँ।
जितनी ही फैलती देश में भीति युद्ध की,
मैं उतना ही कण्ठ फाड़, कुछ और जोर से,
चिल्लाता, चीखता, युद्ध के अन्ध गीत गाता हूँ।

किन्तु, हृदय से जब भी कोई आग उमड़कर
चट्टानों की वज्र-मधुर रागिनी
कण्ठ-स्वर में भरने आती है,
ताप और आलोक, जहाँ दोनों बसते आये थे,

वहाँ दहकते अंगारे केवल धरने आती है;
तभी प्राण के किसी निभृत होने से
कहता है कोई, माना, विस्फोट नहीं यह व्यर्थ है,
किन्तु, बुलाने को जिस को तू गरज रहा है,
उसे पास लाने में केवल गर्जन नहीं समर्थ है।

रोष, घोष, स्वर नहीं, मौन शूरता मनुज का धन है।
और शूरता नहीं मात्र अंगार,
शूरता नहीं मात्र रण में प्रकोप से धुँधुआती तलवार;
शूरता स्वस्थ जाति का चिर-अनिद्र, जाग्रत स्वभाव,
शूरत्व मृत्यु के वरने का निर्भीक भाव;
शूरत्व त्याग; शूरता बुद्धि की प्रखर आग;
शूरत्व मनुज का द्विधा-मुक्त चिन्तन है।

विजय-केतु गाड़ते वीर जिस गगनजयी चोटी पर,
पहले वह मन की उमंग के बीच चढ़ी जाती है।
विद्युत् बन छूटती समर में जो कृपाण लोहे की,
भट्ठी में पीछे, विचार में प्रथम गढ़ी जाती है।

आँख खोलकर देख, बड़ी से बड़ी सिद्धि का
कारण केवल एक अंश तलवार है;
तीन अंश उस का निमित्त संकल्प-शुद्धि है,
आशा है, साहस है, शुद्ध विचार है।

सोच, कहाँ है उस दुरन्त,
पापिनी बुद्धि का मूल, तुझे जो

बार-बार आकर अपनी छलना से छल जाती है।
बार-बार तू उदय-शृंग पर चढ़ क्यों गिर जाता है ?
बार-बार कर में आकर क्यों सिद्धि निकल जाती है ?

ओ विराग को सकल सुकृत का मर्म समझने वाले !
आत्मघात को उच्च धर्म के हित अर्पित बलिदान,
शत्रु के रक्त-पान को
मानवता का पतन, कलुश का कर्म समझने वाले !

ओ निरग्नि ! ओ शान्त ! प्रश्न तेरा गम्भीर, गहन है।
रोष, घोष, हुंकार, गर्जनों से उद्धार न होगा।
भुजा नहीं बलहीन,
रक्त की आभा नहीं मलीन,
अरे, ओ नर पवित्र ! प्राचीन !
दीन, लेकिन, तेरा चिन्तन है।

विजय चाहता है, सचमुच,
तू अगर विषैले नाग पर,
तो कहता हूँ, सुन,
दिल में जो आग लगी है,
उसे बुद्धि में घोल,
उठाकर ले जा उसे दिमाग़ पर।

तुझ से जो माँगते उबलते गीत अनल के,
पूछ कि वे कूटस्थ आग लेकर क्या भला करेंगे ?
क्या प्रमाण है, यह सूखी बारूद नहीं सीलेगी ?
घर में बिखरी हुई बर्फ वे कहाँ समेट धरेंगे।

अच्छा है, वे लड़ें नहीं, जिनके जीवन में
विचिकित्सा जीवित है धर्म-अधर्म की।
अच्छा है, वे अड़ें आन पर नहीं,
न खेलें कभी जान पर,

चबा रही है जिन्हें युगों से
दुविधा कर्म-अकर्म की।

क्योंकि युद्ध में जीत कभी भी उसे नहीं मिलती है,
प्रज्ञा जिस की विकल
द्विधा-कुण्ठित कृपाण की धार है,
परम धर्म पर टिकने की सामर्थ्य नहीं है,
और न आपद्धर्म जिसे स्वीकार है।

तुझ से जो माँगते उबलते गीत अनल के,
पूछ, धर्म की वे किंचित् सीमा स्वीकार करेंगे ?
मानवीय मूल्यों की जब कुछ आहुतियाँ पड़ती हों,
रोएँगे तो नहीं ? पाप से तो वे नहीं डरेंगे ?

अगर कहे तू, युद्ध पुष्प, बमबाजी, फुलझड़ियाँ हैं,
ये रोने की नहीं, मस्त, खुश होने की घड़ियाँ हैं;
दाँतों से तरजनी दबा वे चुप तो नहीं रहेंगे ?
तुझ को वे दानव या दीवाना तो नहीं कहेंगे ?
तब भी श्येन-धर्म ही सच है, गलत युद्ध में पिक है,
पूर्ण चेतना गलत, आज पागलपन स्वाभाविक है।

जूझ वीरता से, प्रचण्डता से बलिष्ठ तन, मन से;
आँख मूँदकर जूझ अन्ध निर्दयता, पागलपन से।

समर पाप साकार, समर क्रीड़ा है पागलपन की,
सभी द्विधाएँ व्यर्थ समर में साध्य और साधन की।

एक वस्तु है ग्राह्य युद्ध में,
और सभी कुछ देय है;
पुण्य हो कि हो पाप,
जीत केवल दोनों का ध्येय है।

सच है छल की विजय, अन्त तक,
विजय नहीं, अभिशाप है।
किन्तु, भूल मत, और पाप जितने घातक हों,
समर हारने से बढ़कर घातक न दूसरा पाप है।

पुरानी और नयी कविताएँ

"नो, हिज़ फ़र्स्ट वर्क वाज़ द वेस्ट।"

—एज़रा पौण्ड

दोस्त मेरी पुरानी ही कविताएँ पसन्द करते हैं;
दोस्त, और ख़ास कर, औरतें।

पुरानी कविताओं में रस है, उमंग है;
जीवन की राह वहाँ सीधी, बे-कटीली हैं;
सरिताएँ जितनी हैं, फूलों की छाँह में हैं;
सागर में नीलिमा है, चंचल तरंग है।

पुरुष बड़े ही पुरज़ोर हैं;
या तो बड़े कोमल हैं, अथवा कठोर हैं।
क्रोध में कभी जो नर-नाहर ये बोलते हैं;
भूमि काँपती है, कोल-कमठ कलमल होते,
दिग्गज दहाड़ते, समस्त शैल डोलते हैं।

नारियाँ बड़ी ही अनमोल हैं;
नख-सिख तक नपी-तुली,
ठीक-ठीक साँचे में ढली हुई;
चन्दन, कदम्ब और कदली की छाया में
दूध और घी पर पली हुई।

भंगिमा स्वरूप को सँवारती है।
वृत्त की गोलाई, जो भी देखे, उसे मारती है।

तेज़ी है अनोखी काम-बाण में।
घाव जो लगेंगे कभी प्राण में,
रेखा में कहूँ तो 'राय-जामिनी' की तूंलिका की
चित्रकारी के वे प्रतिमान होंगे।
छन्द में कहूँ तो रोला-छप्पय के समान होंगे।

चर्म को न छीलता, न छाँटता है।
काम का पुराना बाण
गोदता नहीं है प्राण,
दोहों के समान नपे-तुले व्रण काटता है।

किन्तु, नयी कविता ? गणेश जी का नाम लो।
बुद्धि और कल्पना के चौक पै खड़ी हुई
कहती है, बुद्धि ही कशा है, इसे तेज़ रखो,
कल्पना बढ़े जो तो लगाम ज़रा थाम लो।

कविता न गर्जन, न सूक्ति है।
वीर का न घोष, न तो वाणी खर चिन्तकों की,
चौंके हुए आदमी की उक्ति है।

कविता न पूर्ति है, न माँग है।
सीढ़ियाँ नहीं हैं कि हरेक पाँव सीधा पड़े।
'लॉजिक' नहीं है, य' छलाँग है।

अर्थ नहीं, काव्य शब्द-योग है।
वासना का कीर्तन नहीं है, खुद वासना है,
रागों का यह कागजी बखान नहीं, भोग है।

तन्तुओं के जाल शब्द को जो कहीं बाँधते हों,
सारे बन्धनों के तार तोड़ दो;
अर्थ से बचो कि अर्थ बेड़ी है परम्परा की,
अर्थ को दबाने से ही शब्द बड़ा होता है।
निश्चित-अनिश्चित का संगम जहाँ है सूक्ष्म,
कविता का सद्म निरालम्ब खड़ा होता है।

और वे तरंगमयी नारियाँ ?
पुष्ट देहवाली सुकुमारियाँ ?
सोची गयीं इतनी कि सोच में समा गयीं।
स्थूल से निकल सूक्ष्म कल्पना में छा गयीं।
नारी अब स्वप्न है, अब विचार है।
बाहु-पाश में जो कभी दामिनी-सी नाचती थी
'साइक' में करती विहार है।

नारी शक्ति, नारी धूप-छाँव है।
जानना हो विश्व को तो नारियों के प्राण पढ़ो,
भागना हो विश्व से तो नारी तेज़ नाव है।

और नर भी न नर ठेठ है।

शंकित, सजग, स्याद्वादी अनेकान्तवादी,
कोई 'फ़ास्ट' कोई 'हेमलेट' है।

आखिर, मनुष्य और क्या करे ?

जितना ही ज्यादा हम जानते हैं,
लगता है, आप अपने को उतना ही कम,
उतना ही कम पहचानते हैं।

जितनी ही झाँकी बुद्धि लाती दूर, पार की,
उतने ही ज़ोर से गुफाएँ बन्द गूँजती हैं,
चीख़ती है कुंजी अनजाने, बन्द द्वार की।

केवल कवित्व ही समर्थ है।
सीढ़ियाँ नहीं हैं जहाँ, सारा तर्क व्यर्थ है।
तब भी समस्या बड़ी गूढ़ है।
हम दोनों में से, राम जाने, कौन मूढ़ है।

भूले भी न मेरी विपदाएँ थाहते हैं दोस्त,
केवल पुरानी कविताएँ चाहते हैं दोस्त,
दोस्त और, ख़ासकर, औरतें।

बादलों की फटन

बादलों के बीच छोटी-सी फटन है,
ज्यों, गुलाबी रंग का परदा कहीं से फट गया हो।
इस फटन के बीच से होकर अगर मैं कूद जाऊँ,
कौन गिरने से उधर मुझ को सँभालेगा ?

मूढ़ ! ये पत्थर नहीं, हैं फूल पानी के, धुओं के,
उस तरफ़ का फर्श भी हलका, मुलायम है;
कूदना हो तो अभी ही कूद जा ऊपर घटा में,
अन्यथा, क्षण में, फटन यह बन्द होती है।

आज शाम को

आज शाम को फिर तुम आये उतर कहीं से मन में
बहुत देर कर आनेवाले मनमौजी पाहुन-से।
लगा, प्राप्त कर तुम्हें गया भर सूनापन कमरे का,
गमक उठा एकान्त सुवासित कबरी के फूलों से।

लेकिन, सब को कौन ख़बर दे आया शुभागमन की
फुनगी उठा लता वातायन पर से झाँक रही है,
बारम्बार पवन आता है परदे हटा-हटाकर।
और सितारे अन्तरिक्ष की चौखट लाँघ रहे हैं,
मानो, नभ से उतर यहाँ वे भी आनेवाले हों।

धन्यवाद, मैं, सचमुच, नहीं मरूँगा

प्रथम-प्रथम तुम से जब आँखें चार हुईं,
दहल उठा था मैं अपनी दीनता देखकर।

तुम सुवर्ण साकार,
ताम्र का मैं सिक्का खोटा-सा;
मैं करील श्रीहीन,
दीपती-सी तुम लता विभा की,
तुम गिरि की निर्झरी,
गाँव का मैं सोता छोटा-सा।

कुसुम-कलेवर में प्रकाश का मन देखा,
कहूँ, चमेली को करते चिन्तन देखा।
और सोचने लगा, आरती अब की कहाँ सजाऊँ मैं ?
मन की किस ऊँची तरंग पर इस छवि को ले जाऊँ मैं ?

तुलसी-कानन दूर, निकट पाटल का वन है।
निकट भूमि की गन्ध, दूर निर्मेघ गगन है।
हाय, सूक्ष्म होने पर भी कितना दुर्बल चिन्तन है।

द्विधामूढ़ मेरा मन मुझ को मार गया।
स्वयं आप अपने से ही मैं हार गया।

जो भी करो उपाय, नहीं रुकता है ज्वार रुधिर का।
फूलों की माधुरी बुद्धि को व्याकुल कर देती है।

रोको चाहे लाख, किन्तु वारुणी हृदय की
अपनी लाल तरंग उठा आँखों में भर देती है।

सब समेटकर, तब भी, मैं चुपचाप रहा,
भीतर ही सहता भीतर का ताप रहा।
जैसे अगणित लोग जगत् में मौन रहा करते हैं,
छिपे-छिपे प्राणों की सारी आग सहा करते हैं;

तब सहसा तुम बोल उठीं (है याद ?) एक दिन—
"नहीं, अशिव मरघट की मैं चिनगारी नहीं चुनूँगी,
जितने भी हैं गीत तुम्हारे, सब मुझ को भाते हैं;
किन्तु मरण के गान, हाय, मैं कवि से नहीं सुनूँगी।"

स्वर मेरा काँपने लगा आनन्द से।
खिलने-सा कुछ लगा रुधिर में; और कण्ठ से।
छन्द फूटने लगे, खुशी में भर, पागल बेछन्द-से।

कितनी अच्छी बात ! अभी जीवित है स्रोत सुधा का;
कमी नहीं अब भी करुणा की वायु की,
कवि ! समाज अब भी तुझ को अपना बेटा कहता है।
मातृजाति को अब भी है चिन्ता तेरी परमायु की।

अमृत-मन्त्र; आभास हुआ, मानो, मस्तक पर
अंचल की छाया देकर जीवनदायिनी खड़ी हो।
लगा रही हो स्वयं जिन्दगी, मानो, मुझे हृदय से,
मानो, मुझ पर अनायास हरियाली बरस पड़ी हो।

देवि ! अमृत से तुम ने मुझ को सींच दिया।

अनायास ही तुमने मेरी बाँह पकड़कर
सचमुच ही, कुछ दूर मरण से खींच लिया।

मातृजाति का यही मूक वरदान; नयन से—
जाने, वे क्या अमृत पिला देती हैं।
दृष्टि मात्र से भर देतीं झंकार रुधिर में,
मरती हुई शिखा को देकर स्नेह जिला देती हैं।

धन्यवाद ! मैं, सचमुच, नहीं मरूँगा।
जब तक मेरे लिए कहीं भी करुणा का लवलेश है,
जब तक मेरे लिए सुधा का बिन्दु कहीं भी शेष है,
मरण-कोट पर चढ़ जीवन की ध्वजा उड़ाकर
मैं मृत्युंजय कवि प्रकाश की जय का गान करूँगा।

समाधान

दिनकर केवल प्रश्न उठाता है।
समाधान की बात न कोई पास,
मढ़ समस्याओं का केवल दास।
न तो समझता शास्त्र, न तो क़ानून;
सिर पर चन्दन नहीं, न तो दाढ़ी चेहरे पर,
न तो गले में 'टाइ', न तो नीचे कोई पतलून।
दिनकर केवल प्रश्न उठाता है।
वृथा चौंकते हो,
इस को क्या आता-जाता है ?

उत्तर के गाहको ! निराशा से घबराकर
मरो नहीं, चिन्तनाशील यह देश
नाप में बहुत बड़ा है।
देखो जहाँ, वहीं—
दफ़्तर, कॉलेज-कक्ष में,
सभा मंच पर या विद्या के चबूतरे पर—
समाधान देनेवाला निश्चिन्त खड़ा है।

क्या कहते हो,
प्रश्न गूढ़ हों तो चिन्ता में,
यह मनुष्य भी चार मिनट तक मौन रहेगा ?
राम कहो,

ऐसा-वैसा यह नहीं,
बड़ा पक्का पण्डित है।
इसमें जाकर शमित हो गया
सारा बुद्धि-विवाद है;

बड़े-बड़े ग्रन्थों में जो कुछ बात लिखी है,
इस पण्डित को सभी ज़बानी याद है।

एक मिनट भी नहीं वीर यह मौन रहेगा।
जैसे ही पूछोगे, बेधड़क कहेगा

"हिंसा ? हिंसा कौन बड़ा भारी सवाल है ?
विनय मित्र, मानवता का अभिमान काल है।
हम में जो है क्रोध,
वही सारे जग का अभिशाप है।
गाँधी जी कह गये
कि मरना पुण्य, मारना पाप है।

"और काम का समाधान ?
अत्यन्त सरल है।
पूजा भर है अमृत,
स्पर्श-भावना गरल है।
ब्याह करो, लेकिन रखो,
साबित तब भी ईमान को।
गाँधी जी कह गये,
जन्म लेने दो मत सन्तान को।"

राम, तुम्हारा नाम

राम, तुम्हारा नाम कण्ठ में रहे,
हृदय, जो कुछ भेजो, वह सहे,
दुःख से त्राण नहीं माँगूँ।

माँगूँ केवल शक्ति दुःख सहने की,
दुर्दिन को भी मान तुम्हारी दया
अकातर ध्यानमग्न रहने की।

देख तुम्हारे मृत्यु-दूत को डरूँ नहीं,
न्योछावर होने में दुविधा करूँ नहीं।

तुम चाहो, दूँ वही,
कृपण हो प्राण नहीं माँगूँ।

देवता और प्रेत

कवि के मन में देवता और प्रेत,
दोनों बसते हैं।
प्रेत जब कलम में घुसकर
कविता लिखता है,
देवता हँसते हैं।

देवता की राह हिंसा नहीं,
अहिंसा की राह है।
वे इन्द्रियों से लड़ते नहीं,
पुचकारकर उन्हें पास बुलाते हैं।
देवता के पास
पीपल की छाया होती है।
वे छाँह में इन्द्रियों को
प्रेम से सुलाते हैं।

लेकिन प्रेत कहता है,
जीवन से युद्ध करो.
मारो, मारो, इन्द्रियों को मारो
और अपने को शुद्ध करो।

आँख, कान, नासिका और त्वचा
क्या, हम ने इसी लिए पायी थीं ?

इन्द्रियाँ क्या केवल
मार खाने को आयी थीं ?

देवता अच्छे हैं
कि कीचड़ में ईंट नहीं फेंकते हैं;
न पंचधुनी जलाकर
अपनी इन्द्रियों को सेंकते हैं।

जूठा पत्ता

रामकृष्ण और रमण
रोग की यातना दोनों ने सही थी।
मगर अपने अन्तिम दिनों में
महर्षि ने एक बात कही थी।

जीवन भोज है,
शरीर केले का पत्ता है।
इस पत्ते पर आदमी
भोजन तो बड़े प्रेम से करता है।
लेकिन खाना ख़त्म होते ही
वह उसे फेंक देता है।

जूठा पत्ता भी कभी कोई
सँभालकर धरता है ?

नीरवता

रात ने चुप्पी साध ली है।
सपने शान्ति में समा गये हैं।
अन्तःकपाट आप से आप
खुलने लगा है।
देवता शायद दरवाज़े पर आ गये हैं।

रक्षा करो देवता !

रक्षा करो ! रक्षा करो !
देवता ! हमारा क्या गुनाह है ?
तीन ककारों में पड़ी दुनिया तबाह है।

संसार तुम ने बहुत सोच-समझ कर रचा है।
मगर, एक-एक से पूछ कर देख लो,
कामिनी, कंचन और कीर्ति से कौन बचा है।

मुनियों ने मेनका को गोद में बिठाया।
राजाओं ने कंगालों का कोष चुराया।

और कीर्ति ?
यह चश्मा तो सर्वत्र उबलता है।
शहीद शहीद की जड़ खोदते हैं।
और एक साधु दूसरे साधु से जलता है।

सच है कि कामिनी काल पाकर छूट जाती है।

लेकिन कंचन और कीर्ति ?
रक्षा करो, रक्षा करो देवता !
जब तक जीवित हैं,
कुछ न कुछ खाना ही पड़ेगा।
जितने अच्छे हैं,
उस से ज़्यादा बताना ही पड़ेगा।

नेमत

कविता भक्ति की लिखूँ
या शृंगार की,
सविता एक ही है,
जो शब्दों में जलता है।

अक्षर चाहे जो भी उगें,
कलम के भीतर
भगवान मौजूद रहते हैं।

दूसरों से मुझे जो कुछ कहना है,
वह बात प्रभु पहले
मुझ से कहते हैं।

करुण काव्य लिखते समय
कवि पीछे रोता है,
भगवान पहले रोते हैं।

क्रोध की कविता
भगवान का भौं चढ़ाना है।

और प्रशंसा नर की करो
या नारी की,

भगवान नाराज़ नहीं,
खुश होते हैं।

कविता सबसे बड़ा तो नहीं,
फिर भी अच्छा वरदान है।

मगर मालिक की अजब शान है।
जिसे भी यह वरदान मिलता है,
उसे जीवन भर पहाड़ ढोना पड़ता है।
एक नेमत के बदले
अनेक नेमतों से हाथ धोना पड़ता है !

कविता और विज्ञान

हम रोमाण्टिक थे;
हवा में महल बनाया करते थे।
चाँद के पास हम ने एक नीड़ बसाया था,
मन बहलाने को हम उस में आया-जाया करते थे।

लेकिन तुम हम से ज़्यादा होशियार होना,
कविता पढ़ने में समय मत खोना।
पढ़ना ही हो, तो बजट के आँकड़े पढ़ो।
वे ज़्यादा सच्चे और ठोस होते हैं।

अगर तुम यह समझते हो
कि तुम केवल शरीर नहीं,
आत्मा भी हो,
तो यह अनुभूति तुम्हें,
तकलीफ में डालेगी।
जो सभ्यता अदृश्य को नहीं मानती,
वह आत्मा को कैसे पालेगी ?

विज्ञान की छड़ी जहाँ तक पहुँची है,
बुद्धि सत्य को वहीं तक मानती है।
मशीनों को लाख समझाओ,
वे आत्मा को नहीं पहचानती हैं।

सांख्यिकी बढ़ती पर है,
दर्शन की शिखा मन्द हुई जाती है।
हवा में बीज बोने वाले हँसी के पात्र हैं,
कवि और रहस्यवादी होने की राह
बन्द हुई जाती है।

परम्परा

परम्परा को अन्धी लाठी से मत पीटो।
उस में बहुत कुछ
जो जीवित है,
जीवन-दायक है,
जैसा भी हो,
ध्वंस से बचा रखने के लायक है।

पानी का छिछला होकर
समतल में दौड़ना,
यह क्रान्ति का नाम है।

लेकिन घाट बाँधकर
पानी को गहरा बनाना,
यह परम्परा का काम है।

परम्परा और क्रान्ति में
संघर्ष चलने दो।
आग लगी है, तो
सूखी टहनियों को जलने दो।
मगर जो टहनियाँ
आज भी कच्ची और हरी हैं,
उन पर तो तरस खाओ।
मेरी एक बात तो तुम मान जाओ।

परम्परा जब लुप्त होती है,
लोगों की आस्था के आधार
टूट जाते हैं।

उखड़े हुए पेड़ों के समान
वे अपनी जड़ों से छूट जाते हैं।
परम्परा जब लुप्त होती है,
लोगों को नींद नहीं आती,

न नशा किये बिना
चैन या कल पड़ती है।
परम्परा जब लुप्त होती है,
सभ्यता अकेलेपन के
दर्द से मरती है।

कलमें लगाना जानते हो,
तो ज़रूर लगाओ,
मगर ऐसे कि फलों में
अपनी मिट्टी का स्वाद रहे।

और यह बात याद रहे
कि परम्परा चीनी नहीं, मधु है।

वह न तो हिन्दू है, न मुस्लिम है,
न द्रविड़ है, न आर्य है,
न परम्परा का हर प्रहरी
पुरी का शंकराचार्य है।

उपदेशक

मुजरिम होकर
मुजरिमों को सुधारने का काम,
यह भी एक स्वाँग है
और यह स्वाँग हम सभी लोग
भरते हैं।

जिन मज़ों से हम दूसरों को
रोकना चाहते हैं।
उन से खुद परहेज़ नहीं करते हैं।

वैतरणी के छींटे किस पर नहीं पड़े हैं ?
किस के दामन पर
फूलों के रस का दाग़ नहीं है ?

सपने की देह को नंगी उँगली से
छूने का लोभ किसमें नहीं जगा है ?
कौन इतना जर्जर है
कि उस में जवानी की आग नहीं है ?

तो उपदेश को,
आओ, हम ईमानदार बनें
और मानवता को डराएँ नहीं,
बल्कि यह कहें :

कि जिस सरोवर का जल पीकर
तुम पछताते हो,
उस तालाब का पानी
हमने भी पिया है।

और जैसे तुम हँस-हँसकर रोते
और रो-रोकर हँसते हो,
इसी तरह हँसी और रुदन से
भरा जीवन हमने भी जिया है।

ग़नीमत है कि हर पापी का भविष्य है,
जैसे हर सन्त का अतीत होता है।
आदमी घबराकर व्यर्थ रोता है।

यह बात दूसरी है
कि कोई समृद्धि में है,
कोई अभाव में है।
मगर जहाँ तक मन का सवाल है,
हम सब एक ही नाम में हैं।

सदियों से हमने तुम्हें धोखा दिया है।
मगर अब हम तुम्हें और नहीं भरमाएँगे।
जहाँ तक पहुँचकर हम रुक गये हैं,
उस से आगे का रास्ता तुम्हें नहीं बताएँगे।

एकोऽहम्

मैं दानव से छोटा नहीं,
न वामन से बड़ा हूँ।
सभी मनुष्य एक ही मनुष्य हैं।
सब के साथ मैं आलिंगन में खड़ा हूँ।

वह जो हार कर बैठ गया,
उसके भीतर मेरी ही हार है।
वह जो जीतकर आ रहा है,
उस की जय में मेरी ही जयजयकार है।

मैं ही दाना डालने वाला बुड्ढा रईस हूँ।
मैं ही वह पक्षी हूँ, जो दाना चुगता है।
बैल की पीठ पर बेरहमी से बेंत मत मारो।
मेरी पीठ पर उस का निशान उगता है।

पत्ते का पीला होना
पूरे वृक्ष का रोग है।

यह मात्र संयोग है
कि एक पत्ता पीला है, बाक़ी हरे हैं।
एक व्यक्ति पातक इसलिए करता है
कि सब के भीतर पाप के भाव भरे हैं।

जहाँ भी पुण्य की वेदी है,
मैं अगरु का धुआँ हूँ,

मण्डप से झूलता हुआ
फूलों का वन्दनवार हूँ

और जो भी पाप कर के लौटा है,
उस के पातक में मैं बराबर का हिस्सेदार हूँ।

एक उपकारी सब के गले का हार है।
और जिस ने मारा या जो मारा गया है
उन में से हर एक हत्यारा है,
हर एक हत्या का शिकार है।

भगवान के डाकिये

पक्षी और बादल,
ये भगवान के डाकिये हैं,
जो एक महादेश से
दूसरे महादेश को जाते हैं।

हम तो समझ नहीं पाते हैं,
मगर उनकी लायी चिट्ठियाँ
पेड़, पौधे, पानी और पहाड़
बाँचते हैं।

हम तो केवल यह आँकते हैं
कि एक देश की धरती
दूसरे देश को सुगन्ध भेजती है।
और वह सौरभ हवा में तैरते हुए
पक्षियों की पाँखों पर तिरता है।

और एक देश का भाप
दूसरे देश में पानी
बनकर गिरता है।

कला

जो व्यक्ति लिखता है,
और जो व्यथा भोगता है,
उन दोनों के बीच
झीनी दीवार खड़ी करो।

मैं यह नहीं कहता
कि मेरी उम्र बड़ी करो।
पर जब तक जियूँ,
वीणा सुर में बोलती रहे।
वह मेरा ही नहीं,
सब का दर्द खोलती रहे।

और मैं अपनी सृष्टि में
उसी प्रकार छिपा रहूँ,
जैसे तुम अपनी
रचना में प्रच्छन्न हो।

ताप सब को महसूस हो,
मगर आग नहीं मिले।
कविता में कवि कहाँ छिपा है,
इस का सुराग नहीं मिले।

यह कलाकार की सच्ची स्थिति है।
और मैंने इसे आज तक निभाया था।

लोग समझते रहे कि
मैं देश का दर्द गाता हूँ।
मगर दर्द मैं ने
अपना ही गाया था।

लेकिन, देखो, अब मेरा कितना बुरा हाल है ?

लगता है, मैं कविता नहीं लिखता,
सरापा दर्द हुआ जाता हूँ।

दीवार उठाने में देर हो,
तो कोई चादर ही फेंको।
प्रभो, भरी सभा में
मैं बेपर्द हुआ जाता हूँ।

मनसूबा

मैं अब भी इस आशा में हूँ
कि तुम कुछ कहोगे।
बात को आधे रास्ते में छोड़कर
चुप नहीं रहोगे।

मुझे तो अभी बहुत कुछ जानना है।
आकाश का विस्तार नापना है,
सागर की गहराई थाहनी है
और मरने से पहले
अपने आप को पहचानना है।

ज़िन्दगी के थपेड़े तो खूब खाये,
मगर मनसूबे अभी मरे नहीं हैं।
पेड़ के सभी पत्ते गरचे हरे नहीं हैं,
मगर डाली में रस है।

और रस की यह धारा
क़ैद से छूटना चाहती है।
टहनी के भीतर बेचैनी है
शायद कोई कोंपल फूटना चाहती है।

मैं रेगिस्तान में खड़ा
रस से भरा पेड़ हूँ,

जी भरकर इस बार भी
फलना चाहता हूँ।

जिस रास्ते पर तुम मुझे यहाँ तक ले आये,
उस पर और भी आगे चलना चाहता हूँ।

मालिक, तुम्हारे इस दीये में
तेल अभी बाक़ी है।
अन्धकार को फाड़ने को
मैं अभी और जलूँगा।

तुम्हारी पताका
इस अध-मंज़िल पर
नहीं रुकेगी।
उसे उठाये हुए
पर्वत के शिखरों पर
मैं अभी और चलूँगा।

गाँधी

देश में जिधर भी जाता हूँ,
उधर ही एक आह्वान सुनता हूँ
"जड़ता को तोड़ने के लिए
भूकम्प लाओ।
घुप्प अँधेरे में फिर
अपनी मशाल जलाओ।
पूरे पहाड़ को हथेली पर उठाकर
पवनकुमार के समान तरजो।
कोई तूफ़ान उठाने को
कवि, गरजो, गरजो, गरजो !"

सोचता हूँ, मैं कब गरजा था ?
जिसे लोग मेरा गर्जन समझते हैं,
वह असल में गाँधी का था,
उस आँधी का था, जिस ने हमें जन्म दिया था।

तब भी हम ने गाँधी के
तूफ़ान को ही देखा,
गाँधी को नहीं।

वे तूफ़ान और गर्जन के
पीछे बसते थे।

सच तो यह है
कि अपनी लीला में
तूफ़ान और गर्जन को
शामिल होते देख
वे हँसते थे।

तूफ़ान मोटी नहीं,
महीन आवाज़ से उठता है।
वह आवाज़
जो मोम के दीप के समान
एकान्त में जलती है,
और बाज़ नहीं,
कबूतर की चाल से चलती है।

गाँधी तूफ़ान के पिता
और बाज़ों के भी बाज़ थे।
क्योंकि वे नीरवता की आवाज़ थे।

पुनर्जन्म

जन्म लेकर दुबारा न जनमो,
तो भीतर की कोठरी काली रहती है।
काग़ज चाहे जितना भी
चिकना लगाओ,
ज़िन्दगी की किताब
ख़ाली की ख़ाली रहती है।
शुक्र है कि इसी जीवन में
मैं अनेक बार जनमा
और अनेक बार मरा हूँ।

तब भी अगर मैं
ताज़ा और हरा हूँ,
तो कारण इस का यह है
कि मेरे हृदय में
राम की खींची हुई
अमृत की रेखा है।

मैं ने हरियाली पी है,
पहाड़ों की गरिमा का
ध्यान किया है,
बच्चे मुझे प्यारे रहे हैं
और वामाओं ने राह चलते हुए
मुझे प्रेम से देखा है।

पर्वत को देखते-देखते
आदमी का नया जन्म होता है।
और तट पर खड़े ध्यानी को
समुद्र नव जीवन देने में समर्थ है।

नर और नारी
जब एक-दूसरे की दृष्टि में
समाते हैं,
उन का नया जन्म होता है।
पुनर्जन्म प्रेम का पहला अर्थ है।

पुनर्जन्म चाहे जितनी बार हो,
हमेशा जीवित रहने से
हमें डरना भी चाहिए।

दोस्ती, बन्धन और लगाव की भी
सीमा होती है।
अपने अतीत के प्रति
हर रोज़ हमें थोड़ा
मरना भी चाहिए।

अवकाशवाली सभ्यता

मैं रात के अँधेरे में
सितारों की ओर देखता हूँ,
जिन की रोशनी भविष्य तक जाती है।

अनागत से मुझे यह खबर आती है
कि चाहे लाख बदल जाए,
मगर भारत भारत रहेगा।

जो ज्योति दुनिया में
बुझी जा रही है,
वह भारत के दाहिने करतल पर जलेगी
यन्त्रों से थकी हुई धरती
उस रोशनी में चलेगी।

साबरमती, पाण्डिचेरी, तिरुवण्ण मलई
और दक्षिणेश्वर,
ये मानवता के आगामी
मूल्य-पीठ होंगे।

जब दुनिया झुलसने लगेगी,
शीतलता की धारा यहीं से जाएगी।

रेगिस्तान में दौड़ती हुई सन्ततियाँ
थकनेवाली हैं।
वे फिर पीपल की छाया में
लौट आएँगी।

आदमी अत्यधिक सुखों के लोभ से ग्रस्त है।
यही लोभ उसे मारेगा।
मनुष्य और किसी से नहीं,
अपने आविष्कार से हारेगा।

गाँधी कहते थे,
अवकाश बुरा है।
आदमी को हर समय
किसी काम में लगाये रहो।
जब अवकाश बढ़ता है,
आदमी की आत्मा ऊँघने लगती है।
उचित है कि ज्यादा समय
उसे करघे पर जगाये रहो।

अवकाशवाली सभ्यता
अब आने ही वाली है।
आदमी खाएगा, पिएगा
और मस्त रहेगा।
अभाव उसे और किसी चीज़ का नहीं,
केवल काम का होगा।
वह सुख तो भोगेगा,
मगर अवकाश से त्रस्त रहेगा।

दुनिया घूमकर
इस निश्चय पर पहुँचेगी
कि सारा भार विज्ञान पर डालना बुरा है।

आदमी को चाहिए कि वह
खुद भी कुछ काम करे।
हाँ, यह अच्छा है
कि काम से थका हुआ आदमी
आराम करे।

सूखा पेड़

सूखा पेड़ ग़म की लकीर है।
उसे देखकर वैराग्य का
भाव जगता है
उदासी को देखकर
आँखें उदास हो जाती हैं।
और मन कुछ सोचने लगता है।

मगर एक लता बौरायी हुई है,
इस सूखे पेड़ को घेरकर–
सिर से पाँव तक छायी हुई है।

वल्लरी पेड़ को ढँककर
अपना रूप दरसाती है,
पेड़ के जीवित होने का
भ्रम उपजाती है।

और यह भ्रम इतना प्रबल है
कि पक्षी उस की माया में फँस गये हैं,
पेड़ को जीवित जानकर
उस के कोटरों में बस गये हैं।

सूखा पेड़ लता के वेष्टन में खड़ा है।
यानी प्रेमी प्रिया के आलिंगन में
निर्जीव पड़ा है।

शोक की सन्तान

हृदय छोटा हो,
तो शोक वहाँ नहीं समाएगा।
और दर्द दस्तक दिये बिना
दरवाज़े से लौट जाएगा।

टीस उसे उठती है,
जिसका भाग्य खुलता है।
वेदना गोद में उठाकर
सब को निहाल नहीं करती,
जिस का पुण्य प्रबल होता है,
वह अपने आँसुओं से धुलता है।

तुम तो नदी की धारा के साथ
दौड़ रहे हो।
उस सुख को कैसे समझोगे,
जो हमें नदी को देखकर मिलता है ?

और वह फूल
तुम्हें कैसे दिखाई देगा,
जो हमारी झिलमिल
अँधियाली में खिलता है ?

हम तुम्हारे लिए महल बनाते हैं।
तुम हमारी कुटिया को
देखकर जलते हो।

युगों से हमारा तुम्हारा
यही सम्बन्ध रहा है।
हम रास्ते में फूल बिछाते हैं,
तुम उन्हें मसलते हुए चलते हो।

दुनिया में चाहे जो भी निज़ाम आए,
तुम पानी की बाढ़ में से
सुखों को छान लोगे।
चाहे हिटलर ही
आसन पर क्यों न बैठ जाए,
तुम उसे अपना आराध्य
मान लोगे।

मगर हम ?
तुम जी रहे हो,
हम जीने की इच्छा को तोल रहे हैं।
आयु तेज़ी से भागी जाती है
और हम अँधेरे में
जीवन का अर्थ टटोल रहे हैं।

असल में हम कवि नहीं,
शोक की सन्तान हैं।
हम गीत नहीं बनाते,
पंक्तियों में वेदना के
शिशुओं को जनते हैं।

झरने का कलकल,
पत्तों का मर्मर
और फूलों की गुपचुप आवाज़,
ये ग़रीब की आह से बनते हैं।

परिशिष्ट

दिनकर साहित्य

काव्य

1.	प्रणभंग	1929
2.	रेणुका	1935
3.	हुंकार	1938
4.	रसवन्ती	1939
5.	द्वन्द्वगीत	1940
6.	कुरुक्षेत्र	1946
7.	सामधेनी	1947
8.	बापू	1947
9.	धूप-छाँह	1947
10.	इतिहास के आँसू	1951
11.	धूप और धुआँ	1951
12.	मिर्च का मजा	1951
13.	रश्मिरथी	1952
14.	दिल्ली	1954
15.	नीम के पत्ते	1954
16.	सूरज का ब्याह	1955
17.	नील कुसुम	1955
18.	चक्रवाल	1956
19.	कविश्री	1957
20.	सीपी और शंख	1957
21.	नये सुभाषित	1957

22. लोकप्रिय कवि दिनकर 1960
23. उर्वशी 1961
24. परशुराम की प्रतीक्षा 1963
25. कोयला और कवित्व 1964
26. मृत्ति तिलक 1964
27. दिनकर की सूक्तियाँ 1964
28. आत्मा की आँखें 1964
29. हारे को हरिनाम 1970
30. दिनकर के गीत 1973

गद्य

31. मिट्टी की ओर 1946
32. चित्तौड़ का साका 1948
33. अर्धनारीश्वर 1952
34. रेती के फूल 1954
35. हमारी सांस्कृतिक एकता 1955
36. भारत की सांस्कृतिक कहानी 1955
37. राष्ट्रभाषा और राष्ट्रीय एकता 1955
38. उजली आग 1956
39. संस्कृति के चार अध्याय 1956
40. काव्य की भूमिका 1958
41. पन्त, प्रसाद और मैथिलीशरण 1958
42. वेणु वन 1958
43. धर्म, नैतिकता और विज्ञान 1959
44. वट-पीपल 1961
45. लोकदेव नेहरू 1965
46. शुद्ध कविता की खोज 1966
47. साहित्यमुखी 1968
48. हे राम ! 1968
49. संस्मरण और श्रद्धांजलियाँ 1970
50. मेरी यात्राएँ 1971

51.	भारतीय एकता	1971
52.	दिनकर की डायरी	1973
53.	चेतना की शिला	1973
54.	विवाह की मुसीबतें	1973
55.	आधुनिक बोध	1973

❑❑

णाणं पयासयं

वीर नि.सं. 2470

भारतीय ज्ञानपीठ

ई.सन 1944